DECISIONS

Based on Science

by

Vincent Campbell

Jocelyn Lofstrom

Brian Jerome

National Science Teachers Association
Arlington, VA

Developed with funding from the American Petroleum Institute.

About the Cover

The cover art represents the process of making decisions based on science. The art is inspired by the works of M.C. Escher (1898–1972) who was known for his spatial illusions, impossible buildings, and repeating geometric patterns (tessellations).

Instead of a metamorphosis from caterpillar to butterfly, this is a conceptual transformation. The abstract diamond symbols are changed by adding information; they become something more realistic, but with more detail: a butterfly. Similarly, decision making is a transformation from the simplistic first expression of a decision (*i.e.*, yes or no) to the complexities of scientific evidence and competing interests among constituencies. Once the details of science are incorporated, the decision "comes alive" in all its complexity. Icons representing this process are used throughout the book to illustrate the decision-making steps.

Watercolor butterfly and computer-enhanced transformations by Linda Whynman.
Design by Graves Fowler Associates, Inc.

Library of Congress Catalog Card Number: 97-068383

ISBN: 0-87355-165-6

NSTA Stock Number: PB141X

Printed in the U.S.A. by Kirby Lithographic Company, Inc.

Acknowledgments

We would like to thank the following individuals for their graciously given time and expertise in the development of this project.

American Petroleum Institute

Barbara Moldauer
James Vail, Ph.D.
G.A. Van Gelder, Ph.D.

Curriculum Reviewers

Louise Belnay
Hodgkins Middle School
Westminster, CO

Eugene Bierly, Ph.D.
American Geophysical Union
Washington, DC

William Brune, Ph.D.
Penn State University
State College, PA

Laura Kolb
Aycock Elementary School
Henderson, NC

William George
Georgetown Day School
Washington, DC

Jeffrey Jay, Ph.D.
Northern State University
Aberdeen, SD

Lois Range
Davis High School
Houston, TX

Diane Schranck
Jack Yates High School
Houston, TX

Decision Analysis Reviewers

Rex Brown, Ph.D.
George Mason University
Fairfax, VA

Dennis Buede, Ph.D.
George Mason University
Fairfax, VA

Field Testers

Judy Baldridge
Douglass High School
Douglass, KS

Michael Bannon
Brentwood High School
Brentwood, NY

Donna Carpinelli
Clifton High School
Clifton, NY

Alonda Droege
Steilacoom High School
Steilacoom, WA

Christopher Emery
Amherst Regional High School
Amherst, MA

John Kounas
Westwood High School
Sloan, IA

Diane Schranck
Jack Yates High School
Houston, TX

John Whitsett
L.P. Goodrich High School
Fond du Lac, WI

Art and Design

Illustrations:
   Linda Whynman

Roller coaster diagram (page 70):
   Sergey Ivanov

Design:
   Graves Fowler Associates, Inc.

Layout:
   Christina Frasch

National Science Teachers Association

Decisions—Based on Science is published by the National Science Teachers Association: Gerry Wheeler, Executive Director; Phyllis Marcuccio, Associate Executive Director for Publications. The production of *Decisions—Based on Science* was a team effort by the NSTA Special Publications staff: Shirley Watt Ireton (Director), Chris Findlay and Doug Messier (Associate Editors), Jocelyn Lofstrom (Assistant Editor), Eric Knaub (Program Assistant), and Christina Frasch (Editorial Assistant).

Jocelyn Lofstrom was the project editor. Doug Messier wrote "Recycling" and "Humans vs. Robots in Space." Christina Frasch suggested the format for and revised Part Three. Chris Findlay revised "Marine Resources" and "Old-Growth Forests." Shirley Watt Ireton revised "Ozone," "Groundwater," "Roller Coasters," "Severe Weather," and "Floodplains." The entire staff generously provided the creativity and comments to develop *Decisions—Based on Science* from a good idea into a valuable resource for educators.

Table of Contents

Introduction

Why a Science Book on Decision Making?

The cacophony of modern life can overwhelm us. As good citizens, we must take stands on a wide range of scientific issues. Our personal decisions depend on understanding technology. Not only must we be well informed, but we need ways to make sense of conflicting information. Making sense of this cacophony is a skill called decision making.

Nothing is more important than educating the next generation of decision makers. This book introduces your students to skills and issues for practicing making decisions. The first part of the book provides an overview of the process and gives teaching ideas. The second and third parts contain the decisions, the case studies on how decisions affect our lives.

Recognizing the need for decision-making skills, the *National Science Education Standards* "...describe a vision of the scientifically literate person and present criteria for science education that will allow that vision to become reality."[1] To visualize the scientifically literate person, first visualize a musician. A musician acquires skills by learning the notes and concepts of music. Then, the individual musician applies these skills to benefit a group of players. Similarly, science vocabulary and concepts are like notes on the page. The scientist plays the music of each field of inquiry.

But in traditional science education, students infrequently progress past the notes on the page; they play the occasional song in a short laboratory exercise. Your students can move beyond disconnected science to ensemble pieces—combining science concepts and research with social contexts.

The skills of decision making are embedded in new content standards that augment the traditional fields: Science as Inquiry, Science and Technology, Science in Personal and Social Perspectives, and History and Nature of Science. Students learning science this way would not just memorize information, but use scientific thinking to make everyday decisions. By mastering the skills of decision making, students would be able to identify and state a decision problem; identify viable options; research risks and benefits; make a decision based on rational methods; and present the decision coherently and logically. A scientifically literate person makes decisions—based on science.

For students who learn the logic of decision making, studies reported in the news, labs they conduct in class, symbols and vocabulary words are no longer disconnected pieces but, when combined, comprise the entirety of the scientific process. Students will see that scientists don't do science just for the particular answer to their (usually highly specialized) research-of-the-moment, but to advance toward something more encompassing, a larger question which needs resolution.

By teaching decision making, educators will teach science as symphony. A decision maker conducts the instruments of science to fully play out its music—science's impact on the world.

...Americans are confronted increasingly with questions in their lives that require scientific information and scientific ways of thinking for informed decision making.

National Science Education Standards

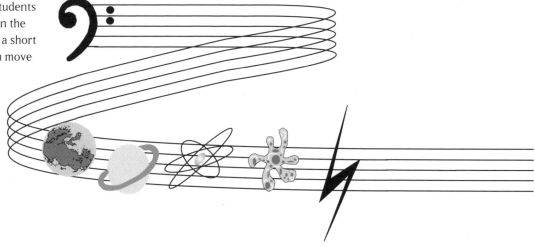

Science is a social enterprise

As students learn science, they read about famous scientists who have discovered something. Most scientists, however, do not necessarily look for something new, in the sense of finding something undiscovered. Instead, the majority of scientific inquiry is directed at quantifying and explaining systems that science has already discovered.

Increasingly, the needs of government and private industry define the focus of science. The scientist who is unaffiliated and self-funded is, for better or worse, a rarity. Science is expensive and needs the contributions of society—its natural resources, its funding, and its public interest.

A society makes decisions regarding how much of its resources to devote to science. While individuals conduct science, society decides what science to conduct. This decision is based on economics, politics, and ethics, along with technological limitations inherent to specific problems.

Of course, scientists determine much of the course of science by their own interests and by lobbying for funds. Research influences social issues and social issues affect research priorities.

Good decision makers are neccessary to this mix because scientific information in and of itself doesn't resolve all dilemmas. We have all been frustrated by conflicting scientific reports. Milk is bad for us; milk is good for us. Earth's atmosphere is warming; it is cooling. We need a logical method, a "scientific way of thinking," to order facts into a coherent summary of a given situation.

Science can't answer all questions because it doesn't exist in a vacuum, without input from those who conduct it and for whom it is conducted. There is always a social context in which science takes place. Science can and does provide guidelines for establishing methods of inquiry, gathering and using information, and manipulating physical elements, but it is generally not up to scientists to decide how individuals and societies use science-based questions, information, and elements. Exploring space, for example, was a social decision made about the use of science and technology. But, while scientists were certainly involved in gathering the information that supported the decision to explore space, it was a decision ultimately made by policy makers.

Another reason that science isn't sufficient is that the answers to some questions are value based, regardless of how scientists have framed them or gathered supporting information. Scientists tell us that flu immunization shots prevent sickness and that too many sulfur oxides in the air cause acid rain. But whether or not to get a shot is a personal decision and involves personal values. How many sulfur oxides to allow in the air is a social decision, and while probably supported by scientific evidence, must consider social values. Scientists, as a rule, don't make our personal and societal decisions for us.

This is not to say that science hasn't affected values over time. Scientific evidence has changed people's opinions about many issues and technologies. But sometimes decisions still come down to values: How much are we willing to pay for pollution controls? How much control do we want the government to have over us in the name of preventive health care?

Societal challenges often inspire questions for scientific research, and social priorities often influence research priorities through the availability of funding for research.

National Science Education Standards

Government, science, and decision making

Science and government generally are taught as separate subjects in schools, as though they belong to separate spheres of life. But science influences government in important ways. This influence is not abstract, but how governments actually function.

A risk assessment is a scientific study of an issue that quantifies the risks of a proposed action to human health and to the environment, and to economic and social systems. The U.S. Federal Government has mandated risk assessments for many major policy decisions in recent years. In 1996, the President issued an Executive Order on Regulatory Planning and Review which mandated "...risk-based priority setting and the balancing of risk reduction and costs in the broader context of regulatory decision making."[2] The U.S. Congress has also called for "...detailed risk assessments and benefit-cost analyses as a basis for determining appropriate regulations and standards."[3] For example, the Federal Government requires local governments to commission an Environmental Impact Statement (EIS) before making zoning approvals. An EIS is simply an environmental risk assessment.

The field of risk assessment is growing in academia and private consulting firms. Researchers conduct risk assessments to help someone make a decision. Clearly, the conduct of science influences government.

Concurrently, the *National Science Education Standards* are calling for citizens who can make sense of the issues of our day, who can "...engage intelligently in public discourse and debate about matters of scientific and technological concern."[4] In the past, societies needed citizen-soldiers; today's technological society needs more citizen-scientists.

How to Use this Book

The book is divided into three parts. Part One provides an overview for teachers, including full background on decision making, introductory activities on important concepts, teaching tips, presentation ideas, and a teaching plan. Once you present Part One concepts to your students, they should be able to apply this knowledge to any decision problem.

Part Two contains Guided Activities that are fully developed with separate student and teacher pages. These activities contain enough information for the students to begin working on the problem. The student pages should be reproduced and handed out. The teacher pages contain suggestions for solutions—or ways of approaching the problem—background information, and teaching plans. They also refer back to Part One concepts that are pertinent.

Part Three contains Independent Exercises that are meant to suggest additional topics for practicing the skills learned in this book. The example problem is outlined in a basic manner, and students should determine for themselves what kind of information they would need to make a decision.

Parts Two and Three will require some preparation on your part. You will need to be familiar enough with the issues to answer questions, and guide students to resources that provide answers. The activities and exercises are extensions of subjects covered in life, physical, or Earth sciences. The course textbook should also be a resource. See Appendix A: Curriculum Matrices to connect the activities to science content standards.

For all these decision activities, it must be stressed that there is no "answer." Most issues contain ethical and economic aspects which will necessarily determine how individual students or classrooms decide. The goal is to teach thinking, not to give answers to questions. The value of these activities is that students can approach problems in a scientifically literate fashion.

How to get started

Teacher preparation
Read through Part One (Background and Curriculum) and familiarize yourself with the concepts and techniques of decision making.

Student introduction
Use the suggested teaching plan at the end of Part One to introduce the topic to your students; or develop your own teaching plan using the teaching tips and the three activities. An introduction to the topic with your students should take two or three 50-minute sessions.

Skills development
After students are comfortable with the Part One concepts, select activities from Part Two as they intersect thematically with your existing curriculum. Use the activities to expand upon content learning. For example, if your curriculum includes a discussion of ecosystems, the activity, "The Politics of Biodiversity" would help students think through the decisions that must be made by the government on ecosystem conservation.

Future curriculum ideas
Use activity formats in Parts Two and Three to develop your own decision-making teaching plans on any relevant topic. For example, the activity, "A Local Decision" provides a format for students to examine local issues.

Teaching strategies and assessment

Decision-making journal
As students learn decision making, they will learn new thinking skills, brainstorm ideas, make notes on their ideas and research, and record conclusions. Because the process allows for a wide range of answers and approaches, students will need decision-making journals to keep records of their thinking processes. In fact, much of the assessment for this book will be based on students documenting their decisions.

A journal has several benefits. Students will learn note-taking skills; they must listen carefully and interpret information for their own understanding. Digesting information in this manner forces students to keep a good record of the class work. Students will examine what they know and find their own words to express that knowledge.

A journal also keeps student notes and class work information in a single place. The journals can be fancy or simple, from an 8 1/2 × 11" notebook to a loose-leaf binder. You will hand out some photocopies from this book, so student journals should have a method for incorporating them, either by stapling in the page or punching holes (loose-leaf binder).

Have students number their journal pages and make a table of contents so they can find certain notes on procedures or examples when needed. A table of contents will help you find specific assignments when you need to review student work.

For more ideas, see *How to Write to Learn Science* in Selected Resources.

Research

These activities take students beyond the usual goals of science education—developing a facility with science concepts and processes—to using those concepts to make decisions. Research should start with the science textbook, but extend to good science or general encyclopedias and the Internet.

Generally speaking, additional research will be needed on the social science aspects of a topic. Newspapers, news magazines, and other periodicals are appropriate sources.

The bottom line: Students should know where to get some basic, yet up-to-date information about the topic. Some resources are listed at the end of each activity, but use your best judgment. As in the real world, the inability to find supporting data contributes to problems with making decisions. This, in itself, is a valuable lesson for students.

Assessment

As noted above, the decision-making journal will be used to guide your assessment of student work. Some notes on assessment and sample grading rubrics are in Appendix D.

Notes

1. National Research Council. 1996. *National Science Education Standards* (Washington, DC: National Academy Press), 11.

2. Robin Cantor. 1996. "Rethinking Risk Management in the Federal Government." *The Annals of the American Academy of Political and Social Science: Challenges in Risk Assessment and Risk Management* 545 (May), 137.

3. Howard Kunreuther and Paul Slovic. 1996. Preface to *The Annals of the American Academy of Political and Social Science: Challenges in Risk Assessment and Risk Management* 545 (May), 8.

4. National Research Council. 1996. *National Science Education Standards* (Washington, DC: National Academy Press), 1.

PART ONE

Background and Curriculum

DECISION MAKING AND SCIENCE:
You Already Think This Way

Traditionally, teaching decision-making skills has not been integral to teaching science skills. Yet there is an intimate connection between scientific thinking and decision making, especially when it comes to using logical and evidential rules to define problems, form and test hypotheses, and translate results into action. Your students will benefit from learning how to make and apply decisions to their scientific inquiry.

Why do you teach science? Is it to make sure students know the Periodic Table of Elements? To have them memorize the organelles of a cell? Most teachers would agree that the facts of science, while important pieces of knowledge, are only building blocks toward the greater goal of scientific literacy. Scientific literacy is a mode of inquiry in which logical and evidential rules guide one to a conclusion consistent with current and historical scientific thought.

Scientific methodology is rightly careful to try to prevent personal values from affecting conclusions, such as how likely it is that a particular action will lead to a particular outcome. However, a decision about which action is best rightly includes personal

According to the principle of testing conclusions, teaching science should promote discussion. While students may not be able to discuss the molecular weight of carbon, they can debate how or if the level of carbon dioxide affects global warming. As a scientific method of examination, decision making allows students to practice the process of reasoned arguments and logical, evidence-based presentations. Decision making reveals the workings of logic, by organizing logic in a chart on a page.

You will notice an emphasis on mathematical models in the activities in this book. One reason numbers are useful in science is that they clearly communicate results, sometimes better than words. Once unfamiliar methods become familiar, math becomes a way of illustrating a scientific fact or prediction, or a way of combining judgments. You may be able to learn facts without math, but you cannot "do" science without it.

The activities in this book present science-based skills and comtemporary issues for practicing decision making. You and your students will find that some issues cannot be answered in purely scientific terms. Perhaps the most important decision

DECISION CHART I: Choosing a book

	Option A: Buy Book A	Option B: Buy Book B
Goal: Save money	Outcome: Costs $29.95	Outcome: Costs $10.00
Goal: Completely cover topics	Outcome: Covers all topics	Outcome: Missing half the topics

and societal values. A good decision maker integrates those values with probabilities of outcomes, and provides a logical, practical way to choose the best option.

You probably want your students to understand that science is a process. For every conclusion, scientists work to either refute or expand on it. Science is always wide open, and needs inquiring minds to keep the process going. The need for accurate scientific information on which to base decisions fuels the scientific process, both philosophically and financially.

students will face throughout their lives is what we—our society—should do with a certain technology or science discovery.

At first, the terms, charts, and mathematics of decision making may seem unfamiliar. But most educated people already think this way. Decision making provides order to our thoughts, and shows us how to take our thinking further. Here's how you already make decisions using logical steps.

Imagine that you must choose one book for your class. You have some goals (or criteria) for the purchase, such as, "I want the best book (quality) for the lowest price (cost)." You identify two books, each of which fits these two goals. Chart 1 summarizes the decision you must make.

The key question is: How much value do you place on having all topics covered in one textbook, compared to saving $20 on another textbook? This is known as "trading off values," or comparing the importance of different advantages of each option.

A decision must include all the important factors. "How interesting the book will be to students" is a goal not mentioned above. Including this goal in your analysis might change your decision.

Different people can value outcomes differently. For example, you might place more importance on the added 50 percent in completeness than on a $20 difference in cost, whereas a fiscally-minded school

board might do the opposite. This is not to say that one is right and the other wrong, but rather that different individuals or constituencies affected by a decision may have genuinely different values and priorities.

Obviously, most decisions are more complex because they involve more factors to consider than our book example. This requires a method to extend the process. Decision making—as presented in this book—is a rigorous method for evaluating and extending complex decisions.

The specific techniques for making decisions will be presented step by step. They enable you to teach decision making as a qualitative process (no numbers used) and to stop there. If you wish the class to use the same concepts but apply numbers and equations, this can be done either as an enhancement to a given activity or as a separate activity. Teaching Plan: Part One (page 20) contains presentation ideas.

DECISION MAKING:
The Steps

A decision resolves a divergence of paths; each path will have benefits and risks. A good decision maker identifies each path's benefits and risks, uses evidence to weigh them logically, and then decides. The terms used in this book are: decision (divergence of paths), options (each path), and outcomes (a path's benefits and risks). There are several decision-making techniques, and decision makers in business and government use a range of terminology and procedures. But decision-making processes generally follow four steps.

The four icons used throughout this book represent these four steps. They illustrate decisions based on science—using science facts to transform knowledge from simple to complex. The icons are inspired by techniques of the artist M.C. Escher. You may want to explain their meaning to your students because they appear in many places in this book.

Important personal and social decisions are made based on perceptions of benefits and risks.
National Science Education Standards

What's the Decision?

This icon represents the first expression of the decision. At this point, preconceived notions and personal values may interfere with one's ability to comprehend the complexity of the issue.
Action: The decision maker identifies a hazard with a certain risk, considers the effects, and makes a preliminary list of significant factors impacting the decision. See page 6.

What Should Happen?

This icon represents the transformation from identifying the decision abstractly to defining its constituent parts.
Action: The decision maker identifies possible solutions and determines likely outcomes of each option. The decision maker also ensures that all aspects of the decision have been considered. See page 11.

What Do We Know?

This icon represents a greater comprehension of the issue. By incorporating research into the analysis, an image of the whole emerges.
Action: The decision maker researches the probabilities that the identified outcomes will occur. The particular decision defines what information is needed. See page 12.

What's the Answer?

This icon represents the full understanding of the decision in all its complexity. In one sense, the decision has "come alive."
Action: The decision maker uses a rational method of analysis to decide. Nevertheless, the decision maker may identify further research needed. See page 15.

STEP ONE: What's the Decision?

We must clearly identify and state the decision to be made. Just stating the problem can be complex, but it begins the process of ordering our thoughts. Information about emerging problems may be incomplete; part of the decision maker's skill is determining what information is most needed. Therefore, the first question we must ask is:

Whose Decision Is It?

Decisions are made at personal, local, national, and international levels. The decisions a local school board makes are different from the decisions the United Nations makes. Why? Because the scope of influence and the resources of the decision makers are different.

Each decision activity in this book can be analyzed from various perspectives and using different scales. Students should be clear on whose decision it is; they should understand the difference between what we should do as a society and what an individual should do in a certain case. Environmental decisions provide good examples of how both personal and societal values can come into play in the decision-making process.

For example, a decision to recycle newspapers is affected by the scope of the decision. The resources conserved, the energy used for transport, and the labor spent recycling are clearly different for one person than they are for 250 million people. While the societal value may be that we all want to recycle, the actual decision may be at the individual or community level. Identifying whose decision it is avoids confusion and errors in logic.

Consider cigarette smoking. In the past, the choice whether or not to smoke was seen as a purely

> Natural and human-induced hazards present the need for humans to assess potential danger and risk.
>
> *National Science Education Standards*

personal issue. In light of evidence that second-hand smoke can increase the chances of lung cancer, some states and the Federal Government have made a public policy decision to restrict smoking in public buildings and airplanes. The actual decision is still a personal one, but policy has become governmental.

WHOSE DECISION IS IT?

As students consider the activities in this book, have them reword their decision from several different viewpoints. This will show them that, when they do make a decision, the options and outcomes should be consistent with the identified decision maker.

Risk Assessment

We make decisions in response to a *hazard* (see box, below left). To compare the paths or options in a decision, one must evaluate the hazard and calculate the chance of it occurring and how "bad" the result will be. The National Research Council lists many categories of hazards: physical phenomena, chemicals, organisms, commercial products, human behavior, or even information which could adversely affect someone or something. A hazard has a potential for harm: the hazard may or may not happen. The probability a hazard will occur and the extent of harm it might cause is defined as a hazard's *risk*.

Risk assessment, then, is the use of scientific studies to predict risk. Risk assessments must have two components: *exposure* and *intensity of effect*. The exposure indicates whether a certain person, animal, ecosystem, etc., will come into contact with or be affected by a certain hazard. The intensity of effect identifies what might happen at the expected exposure level. A hazard may be very toxic (intensity of effect) but, if the chance of exposure is almost zero, then the risk of the hazard is very slight. Conversely, if you encounter a hazard every day that is not very damaging to your health, the hazard's risk is also slight. For example, a risk assessment for the hazard of rain on a school event would take into account: the probability that it will rain (exposure); and how much it will rain (intensity of effect).

To provide the information needed to assess risk, researchers conduct experiments to measure exposure and intensity of effect. Even if a study is not created to assist a particular decision, it may affect future decisions. For example, the field of epidemiology studies the effects of disease on human populations. These studies report on the connections

HAZARDS AND THEIR RISKS

*The terms used in this book have precise meanings that may be different than everyday use. The term **hazard** means "an act or phenomenon that has the potential to produce harm or other undesirable consequences to humans or what they value."[1] But the hazard may or may not occur, and the extent of harm it may cause varies. So, another term, **risk**, is used to describe the characteristics of a hazard: **exposure** (what is the chance that the hazard will occur?) and **intensity of effect** (what degree of harm will the hazard cause if it does occur?). **Risk assessment** is a scientific process that analyzes a hazard's risk. The **effects** of hazards are what happens when the hazard occurs.*

between exposure to certain substances and the resulting intensity of disease. Decision makers used these studies to legislate on a number of health hazards, including smoking. However, a complete risk assessment considers the effects that a hazard has not just on human health, but on other living beings, and on ecological and social systems.

Human and ecological effects of hazards

The most readily recognized effects are those on ecological systems and human health. Ecological effects include health risks to animals, plants, and ecosystems. Effects should be considered in terms of both current and future hazards. For example, scientists measuring global warming look at past measurements, collect current data, and make predictions about the future.

Human and ecological effects depend on the amount of exposure to a hazard. Have your students keep this in mind as they evaluate hazards. For example, one can be exposed to a hazard (driving a car) every day for years or one can have only one quick exposure (x-rays).

Students should also understand the difference between *chronic* and *acute* effects of hazards. Acute effects are generally fast responses to single exposures of a hazard (food poisoning). Chronic effects happen after repeated exposures (cigarette smoking). Chronic effects are generally more difficult to measure because they take place over time and may be missed in studies.

Social effects of hazards

Social effects are generally not considered part of a science curriculum, but a risk assessment is incomplete without them. Important social effects to consider include economic impacts, effects on government and ethics, and how people perceive the hazard.[2] Social effects should be considered in terms of both present and future effects. Social effects are important to consider because society decides what science to do. If there are social effects to a hazard that scientists and decision makers ignore, they may find themselves defending against public outrage. Despite the fact that social effects may have less "hard" data than human health and ecological effects, they should be measured and then factored into any decision-making process.

Economic effects are the monetary costs of the hazard. Costs are sometimes measured in potential losses. When a manufacturing plant moves into a neighborhood, the value of nearby homes may decrease as people move to other neighborhoods. One economic effect of the plant is financial losses to

INTERDISCIPLINARY EFFECTS

Human and ecological effects cross science disciplines. For example, ecosystem health incorporates subjects like geology, meteorology, physics, and chemistry. Even if you do not have time to cross disciplines in your teaching, let students make the connections abstractly. This will help them see the integrated nature of knowledge.

Also, encourage students to look at a hazard over time: What will happen if this hazard continues? Will the effects be cumulative (i.e., more severe in the future)?

property owners. Costs also can be secondary effects of a hazard. For example, if someone becomes ill from a hazard, that person suffers both health effects and economic effects (*i.e.*, cost of health care, lost time at work).

Government effects include hazards, such as political or social unrest, which affect how a government conducts its business. Government effects also include new laws that must be passed regarding a hazard. The accident at Three Mile Island helped bolster the environmental movement in the United States. The government then increased its emphasis on environmental issues.

Ethical considerations encompass how society chooses to react to a hazard in a moral or values-based manner. For some people, pollution is not just a human health issue, but a moral issue: Polluters should be punished. People's right to make choices regarding their lives is an important consideration. What if a carefully-thought-out decision is overturned in court for violating someone's rights? The public's ethical reaction to a hazard influences what options are available to a decision maker.

ETHICS

Students may find it helpful to think through the ethics of an issue if they consider fairness, prevention, and rights.[3]

◆ *Is the hazard "fair" (equally distributed among the affected population)? For example, in the United States, acid rain affects the Northeast region even though the majority of sources are in the Midwest.*

◆ *Is the hazard preventable? Is there a prevention option?*

◆ *Does the hazard affect people's rights? For example, some environmental laws restrict the use of the habitat of endangered species even on private lands. This restricts the rights of the individual who owns that land. The decision maker may decide after deliberation to restrict use, but the individual's rights should be a consideration during the decision-making process.*

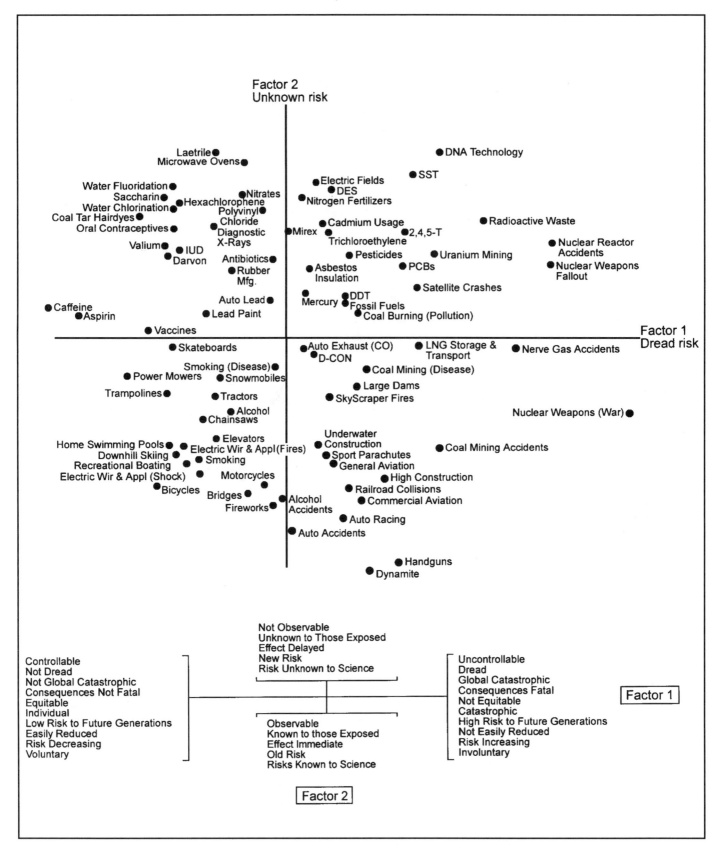

FIGURE 1: Risk Space charts the public perception of hazards based on the hazard's characteristics. The contributing factors are listed at the bottom of the figure.

Reprinted with permission from *Science* 236 (1987): 282 (Paul Slovic, "Public Perception of Risk"). Copyright 1987 American Association for the Advancement of Science.

Public perception of hazards

While professional analysts look at the numerical, measurable risk of a hazard, lay people tend to use several different criteria, including how well the hazard is understood, if it is voluntary, how many people are affected, and the degree of control they can exert to prevent the hazard. *Risk space* is a term used by researchers to describe how the public reacts to different hazards based on the hazards' characteristics. Risk space also refers to a grid developed to chart the public's perception of hazards. (See Figure I.)

RISK SPACE

The activity "Public Perception of Hazards" (page 25) introduces students to the concept of risk space.

The grid helps explain why some hazards are stressed in public debate when other hazards (more common and controllable) are overlooked. Hazards that individuals cannot control (upper right quadrant of grid) are those for which the public wants govern-ment control even if the risk of the hazard is slight. Therefore, risk space illustrates how some hazards which are statistically dangerous (*e.g.*, cause more deaths per year) may be ignored by the public because they are seen as everyday and controllable (car accidents are a good example).

One explanation for why the public reacts this way involves the nature of modern society. The per-ception of risk by the public is greatly influenced by the idea of control over the hazards in one's life. Technology has made life more complex. We depend on many technologies for our daily chores. But most of us have only a general understanding of how these products are made, how they work, and even how to fix them if broken. As one researcher writes: "...we actually know far less today than did our great-great-grandparents about the tools and technologies on which we depend."[4]

If the public perceives that important questions are not answered to their satisfaction, a sense of unease results from submitting to such a hazard (the technology). Therefore, new technologies, like biotechnology, are resisted because the public has significant doubts about whether potential hazards can be identified and handled.

While certain hazards may have equal risk, pre-conceived notions can make some hazards seem more dangerous than others. For example, smoking has a good chance of causing lung cancer. Yet, since it is common, observable, and controllable most students may not rate it as "horrible" as nuclear warfare, which has a small chance of occurring but would produce a much more horrible result. We should be careful when we automatically characterize a hazard without understanding its actual risk. A risk assessment will give us facts to support a characterization.

Once students understand risk assessment, they will learn to ask valuable questions about a hazard. This will help them understand what a hazard is, how scientific information is collected and analyzed, and how results of studies are communicated.

One difficulty in conducting a risk assessment is that information is not always easily available. However, students can benefit from understanding that gathering information is time consuming and sometimes difficult. Encourage students to gather as much information as possible within a certain time frame, and continue with the decision exercise even if some information is missing. These "holes" can be filled in later if you wish. Sometimes we have to proceed without all the pieces of the puzzle.

Identify Goals

The last part of Step One: What's the Decision? is to identify certain important considerations of the decision. These are known as criteria or *goals*. Goals are expressed in terms of how much you value one outcome over another. For example, do you buy a less expensive TV and save money (goal), or spend more for one with better quality (goal)?

After a risk assessment is done for a hazard, the decision maker needs a list of goals. Your students can use their list of effects to pick out goals, or dimensions of value, which pertain to the hazard. But goals need not be a single example from each effect category (*e.g.*, environmental). There may be multiple goals for each effect.

Criterion: a standard, rule, or test on which a judgment or decision can be based.

Webster's Ninth New Collegiate Dictionary

IDENTIFY GOALS

A good way to identify goals is to have students take positions representing different constituencies and role play how each constituency might see the situation. Different constituencies will have different goals. For example, an environmental group will have different goals than a power company. At this point (Step One) students need not evaluate the importance of each goal. All that is required is a debate over whether a goal is pertinent to the decision situation. A formal process for weighting the goals and options will be presented in Step Four.

Make sure students include social effects. Goals for social effects are important because they integrate values into the decision situation. For example, integrate economic aspects in a decision on endangered species by including the goal, "Keep jobs."

Brainstorm goals by thinking of the hazard in terms of best and worst cases. What most terrible thing do you want to avoid? State it as a goal. What's the best thing you hope to have happen from the decision? State it as a goal. Remember: a goal is not an action/ option, but a measure of value for an option to be defined later.

If two goals are similar it may be best to combine them into one. But if a goal includes many different consequences that are hard to compare, it may be best to break it into two or more goals. Use the guidelines below.

Guidelines for combining goals: If the same topic would be counted twice under two different goals, then combine the two goals or narrow one of them so there is no overlap. Otherwise the topic is "double counted" and may have more weight in the decision than it should.

♦ Example: Students might have one goal called "Avoid nuclear radiation" and another called "Protect human health." But one effect of human exposure to radiation is higher incidence of cancer. So, this consequence or outcome would be counted under both goals.

♦ Solution: Simply include radiation effects as part of the "Human health" goal, and do not have a separate goal on radiation.

Divide one goal into many: A goal may be so general that it includes consequences which are hard to compare.

♦ Example: "Reduce environmental effects" as a goal is so broad that it includes outcomes such as eroding marble statues, killing fish, and even human health effects.

♦ Solution: Treat them as separate goals.

Step One Summary

Key concepts

♦ A *decision* resolves a divergence of paths; each path will have benefits and risks.

♦ A *hazard* is an act or phenomenon which has the potential for harm. The potential harm is called a hazard's *risk*.

♦ Risk consists of the *exposure* and the *intensity of effect* of a hazard to a given population, material, or system. *Risk assessment* is the use of scientific studies to predict risk.

♦ Common *effects* of hazards include: human health, ecological, economic, governmental, and ethical. A special kind of effect is *risk space*, which is the public's reaction to a hazard based on its characteristics (controllability and familiarity).

♦ A *goal* is a criterion of value which helps decision makers evaluate options. A goal could be to avoid the risks of a hazard and/or to achieve some positive benefits.

Procedure

(A) Before students do any analysis, have them write in their journals what they think the decision should be. See page xi for information about student journals.

(B) Whose decision is it? Determine at what level (personal, local, national, international) the decision is being made and make sure all goals and options are stated at this level.

(C) Define the hazard and evaluate its risk (intensity of effects and exposure). Consider risk space and human health, ecological, economic, governmental, and ethical effects. Also, consider effects in the present and future. Be as specific as possible, but leave "holes" if a lack of information is holding up the process. Return to the research later; or decide without the information.

(D) Use the risk assessment to make a list of goals (values) for the decision. The list of goals is a preliminary list which can be amended.

STEP TWO: What Should Happen?

Identify Options

Now that we have identified goals for the decision, we must identify options, or alternative courses of action in the decision. Options should meet as many of the goals from Step One as possible, but may not meet all. At first, do not screen out possible options because they don't answer every goal; simply list as many as seem feasible. A good way to brainstorm options is to look at each effect of the hazard and think up ways to solve the problem. Students may find that several effects can be changed by one action. Try not to decide on a course of action before adequately considering how it affects each of the goals.

With a complex problem, there may be many different possible action plans. At the start of the decision process, do not define an option as a combination of actions. Multi-step actions make comparisons difficult, and combining actions can always be a later step after the separate actions are compared and analyzed. In real life, the final choice often does boil down to choosing between two complex strategies, each involving many actions.

Identify Outcomes

The *outcomes* are the consequences an action/ option may have for each goal. This step may seem fairly simple. However, some outcomes are uncertain, so we need to take this into account. Uncertainty and probability are two related concepts we use in such cases.

Uncertainty means that we may not know exactly how an option will impact a goal. For outcomes

IDENTIFY OUTCOMES

Use the following method to identify outcomes. Using a decision chart, enter the list of goals as row headings (on the left), and the most promising options as column headings. In each box of the chart, write a word or phrase summarizing the effect of the option on the goal. If students can't find good scientific information to help, enter a best guess on the outcome. If there is more than one effect on that goal, list them, leaving some space after each effect to make notes later about any uncertainties.

where there is uncertainty, have students make a best guess and note that the outcome needs research. Even professional analysts must deal with a great deal of uncertainty in decision making. For some issues, we may not be able to find an answer. The decision will have to be made with uncertainty.

The probability of an outcome occurring can be difficult to estimate. Social effects, such as job losses or urban migration may be guesses even in best cases. Keep this in mind when filling in probabilities. Step Three covers this topic in greater detail.

IDENTIFY OPTIONS

When students are making a decision, have them list in their journals possible actions to take. Then, they should discuss in groups how each action might affect the goals they have listed. Considering the effects of an option may bring to light a new goal. When appropriate, include "Do nothing" as one option so students will realize the consequences of no action. As a group, have students decide which two to four options they think are most promising. This can be based on a vote of the group if they can't reach a consensus. These are the options they will analyze next using a decision chart.

Organizing Your Thoughts: Decision Charts

Real problems usually don't come apart neatly, even with the best of analyses. A decision chart (see Chart 2, next page) is a tool for analyzing the elements of a decision. Laying out a decision chart with the goals, options, and outcomes helps students see at once all the elements and how they relate to each other. This is especially helpful where there are several different important goals affected by the decision, and tradeoffs to be made on each goal. A decision chart also allows us to add probability and value to the analysis.

A decision chart is simple to construct. List goals down the left side of the chart, and options across the top. As we have seen, an outcome is how a particular option affects a goal. In the decision chart, outcomes are the intersecting boxes between each goal and option.

Suppose you are deciding where to hold a big school event. Your decision chart might include two options, Indoors and Outdoors. A goal probably would be: "Enjoy the event." The outcomes in your decision are the "intersections" between the goal and the options, or how situating the event indoors or outdoors affects your enjoyment. See the activity, "Planning a Big School Event, Part One" for more explanation of goals, options, and outcomes.

	Option X	Option Y	Option Z
Goal A	outcome AX Expected Value(AX) = P×V	outcome AY Expected Value(AY) = P×V	outcome AZ Expected Value(AZ) = P×V
Goal B	outcome BX Expected Value(BX) = P×V	outcome BY Expected Value(BY) = P×V	outcome BZ Expected Value(BZ) = P×V
Goal C	outcome CX Expected Value(CX) = P×V	outcome CY Expected Value(CY) = P×V	outcome CZ Expected Value(CZ) = P×V
Expected value of each option=	**Total for Option X**	**Total for Option Y**	**Total for Option Z**

NOTE: P = Probability, V = Value. Expected value is the product of the outcome's value (V) multiplied by its probability (P). For certain outcomes, P = 100 percent (or 1).

Each outcome box is quantified by calculating its *expected value*, which is the product of the outcome's value (V) multiplied by its probability (P). Value is derived from how well an outcome fulfills a particular goal. If an outcome is certain, P = 100 percent (or 1), and the expected value is V × 1, or V. Using probability in decision charts will be explained in Step Three.

Step Two Summary

Key concepts

♦ *Options* are alternate courses of action.

♦ *Outcomes* are the results of each option with regard to stated goals.

♦ A *decision chart* graphically organizes the elements of a decision.

♦ Outcomes have an *expected value,* which is measured numerically by assigning a *probability* and a *value* and multiplying the two factors.

Procedure

(A) Create a decision chart using the goals from Step One.

(B) Deliberate on and select two to four options.

(C) Write best guesses on outcomes.

(D) Identify initial research needs from outcome boxes that are missing information.

DECISION CHARTS

Decision charts will be explained in greater detail in Steps Three and Four. Use the activity "Planning the Big School Event, Part One" (page 21) to introduce your students to the concepts of probability and value. Students practice using decision charts in "Planning the Big School Event, Part Two" (page 28).

STEP THREE: What Do We Know?

Step Three is gathering information. Of course, students should have done some research to complete the risk assessment. But the information gathered in this step is quite specific; it must fill in the holes in the decision chart. Therefore, this research is not just investigating the topic, but finding the specific information to fit the needs of the decision situation.

For student research, divide the goals so each group researches one goal. Then, make the information available to all groups. If there is not time for research, have them discuss and analyze the outcomes using their own previous class work and experience.

You may give the groups any length of time, from a class period to several days, to search for information on outcomes. Where information sources disagree on an outcome, have the groups take into account the qualifications of the sources (if known) and any bias the sources may have, such as from their employment or financial interests.

Groups should first try to estimate what the outcomes will be using their own knowledge, and any summary information provided by this book. They can construct a draft chart, write in estimates, and replace them with better data after the research. This will give students a realistic experience of what decision makers face when they don't have a budget to do extensive research, or they have to make a decision quickly. It will also help students understand why research is important to decision making.

The Value of Perfect Information

For some of the decision situations students consider, they will be able to fill in a decision chart completely; there will be *perfect information.* Perfect information means probabilities can be reasonably approximated and outcomes predicted. But in scientific research, most information is not perfect. In fact, for many issues information is still missing due to lack of funding for studies or to the difficulty of studying the problem. For example, some problems take place over long time frames (global warming and cooling trends) or on such a huge scale (total global weather patterns) that multiple studies are needed to even begin to learn how a system works.

So, while some activities will have perfect information, students should understand that most decisions are made without it. Policy makers often must make decisions of national interest without the complete information one would assume they need to decide. Science has not failed, it just hasn't caught up to the need. Instead, decision makers use best guesses and apply values to their evaluations. As students develop decision charts for activities in this book, remember that some issues won't have a conclusive answer. This can be frustrating for students (and teachers).

Quantifying values can be an obstacle. Decision makers' values can seldom be measured with high precision. In fact, studies have shown that any one person's values often shift somewhat from week to week, even hour to hour, depending on a person's experiences, mood, and thoughts at a given moment. It is seldom worth trying to estimate a person's values at the level of precision we demand when measuring physical properties such as distance and time. Estimating values will be discussed in Step Four.

What to research: The main goal of research is to quantify outcomes in the decision chart that students have developed.

What not to research: Information about an issue is not necessarily useful information. Don't let students get sidetracked just gathering information. They should limit their research to the identified outcomes.

Stating Probabilities of Outcomes

A *probability* expresses the likelihood of a possible outcome. A large part of science consists of conducting research to identify outcomes and estimate their probabilities. There are two ways to express probability: all-or-nothing events and continuum events.

All-or-nothing events

Some events either happen or don't happen. It either rains today or it doesn't. A candidate for president either wins or loses the election. A meteorite either collides with Earth or it doesn't. With these all-or-nothing events, we usually estimate the probability of the event. For example, the probability of rain today may be 30 percent; the probability of no rain is then 70 percent.

When entering an all-or-nothing event in a decision chart, make two columns in the outcome box: Yes and No. Enter the probabilities and outcomes for each situation as illustrated in Chart 3 (page 14).

Continuum events

Most events of human interest are those that happen to some degree instead of all-or-nothing. Rainfall in a city over a year is not a yes or no event. Some years it rains more than others, some less. You may like music, but you like some types of music more than others.

With events on a continuum, it is common to estimate the *most likely* degree and establish a *confidence interval* around that degree. In statistics, a confidence interval is usually based on an actual distribution of data. In the absence of such data, decision makers often use experts' subjective judgments about the likely range of possible consequences, and may call it a "credible interval" to distinguish it from a "data-based confidence interval." In many decision situations, a confidence interval may mix data and subjective judgment because the data are from a situation somewhat different from that which the decision maker faces. The credible interval is usually interpreted to mean that 95 percent of the time the outcome will fall within that stated range.

For any outcome in a decision you can use either all-or-nothing or continuum methods to express the probability of outcomes. Actual probabilities may be difficult to find. Instead, have students estimate and then select the most uncertain outcomes. Label these "Uncertain." In Step Four, the uncertainty can be taken into account.

For example, in Option B (Chart 3, below), let's suppose there is a drought threatening a farm. Based on expert opinion, you may believe the most likely number of cows that will die is 30. Expert opinion varies and even the experts say they are not sure, but you are almost sure that the number of cows that die will fall between 10 and 60. That is, your 95 percent confidence interval is 10 to 60 cows.

When addressing future events, we need to use the most accurate data available if we expect our estimates to be reasonable. Even if we don't have accurate data, we can state the probability of our belief that an event will happen. For example, suppose you want to estimate how much large scale solar power will cost 20 years from now compared to power generated by burning coal. No one really knows. After studying information on what the present costs are and what new technology is on the horizon, your best estimate may be that solar power will cost around five percent more. Thus, you may feel almost certain that it will cost somewhere between 15 percent less and 20 percent more. These are beliefs based mainly on subjective judgments. In real life, we often have to make decisions without scientific data, or using data from a somewhat different situation. In these cases, we rely in part on our judgment and good sense to estimate the likelihood of an event.

Optional probability step for advanced students

Advanced students may want to integrate probability into their expected value calculations. Instead of assuming the most likely outcomes, students can compute expected value in a more precise manner.

All-or-nothing probabilities

Refer again to Chart 3. Instead of assuming that Option A will not occur because "No" is more likely than "Yes" (.80 versus .20), decision makers use a weighted combination of these two outcomes. They assess the importance of advantages, once assuming the "No" outcome and once assuming the "Yes" outcome. They then average the results giving the "No" outcome four times as much weight as the "Yes" outcome (since .80 = 4 × .20).

The expected outcome for Option A should be as follows: Suppose that, if the option doesn't happen, then 100 cows will die. If the option does happen only 25 cows will die. The *expected value* outcome is: $(.20 \times 25) + (.80 \times 100) = 85$ cows will die.

Continuum probabilities

For uncertain events on a continuum (see Chart 3), look at the 95 percent confidence interval (10–60 cows die). Using the best information available, estimate the most likely point within that continuum. In this case, it is 30. Use this most likely point to judge the value of that outcome. By estimating the most likely point, in effect, you incorporate probability (P) and value (V) by estimating the most likely (P) value (V) within the confidence interval. There is no need to multiply a probability factor because it is incorporated into the estimate. In the example in Chart 3, the expected value outcome is: 30 cows will die.

Although students will not use the confidence intervals to compute expected value, they can use the intervals in sensitivity analysis (page 18).

If students can use equations and estimate probabilities for outcomes, show them how to calculate expected value outcomes. Ask them to apply it to probability for at least one outcome in their charts.

Step Three Summary

Key concepts

♦ *Perfect information* in a decision situation is rare. Many decisions are made with best guesses and approximations. Research should be limited to information pertinent to the decision, not just "noise" information.

♦ *Probability* is a way of expressing the likelihood of an outcome. *All-or-nothing* events are expressed as a percentage chance of the event

DECISION CHART 3: Expressing Probability		Option A (all-or-nothing)		Option B (continuum)
	Goal X	*Yes, P = .20*	*No, P = .80*	
		Result (V): 25 cows die	*Result (V): 100 cows die*	*Result (V): 10–60 cows die (95% confidence)* *Most likely: 30 cows die*

happening (yes) or not happening (no). *Continuum* events are expressed as a range of likely outcomes. *Confidence intervals* use a combination of real data and estimates to state the range within which an outcome is "very likely" to fall (usually 95 percent probability).

Procedure

(A) Research outcomes.

(B) Make probability estimates. Keep in mind that events are either all-or-nothing or continuum. *Note: This step may be omitted if too difficult for some students. Instead, just have students identify any highly uncertain outcomes.*

(C) Complete decision chart with information from research.

STEP FOUR: What's the Answer?

Once you have a reasonable summary of the goals, options, outcomes, and probabilities, you may think the answer should be obvious. But sometimes the amount of data is overwhelming, and you need a logical technique to determine the results. A method called *Decision Analysis* simplifies information to common units of meaning so you can weigh options. The following techniques are simplified versions of techniques used by professional decision makers. See Selected Resources for further reading on Decision Analysis.

Analyzing the Decision: Method One

As a first attempt at analyzing the decision, ignore any uncertainties and leave out the probability estimates. The first question to consider is: For each goal, which option has the better outcome? Each goal is first considered separately. One option has an "advantage" over the other on this goal, in the ordinary sense of the word. Decide which option is best for each goal, and put a (+) in that outcome cell. Put a (-) in the least preferable option for each goal.

Here you can see why quantifying outcomes is very important. It is very difficult to compare outcomes if they are vague descriptions. Nevertheless, in many cases, you will have to make decisions without perfect information.

Thinking carefully about *advantages* (the differences between two outcomes) is one of the most important steps in Decision Analysis. Students should compare advantages of options across goals. They shouldn't

WHICH METHOD?

In Step Two, we discussed how a decision chart helps you analyze a decision (page 11). Once the chart is completed, there are two methods to compute the result. Method One uses "importance bars" drawn by students to represent their evaluations of options. This method is less math intensive, although it requires some explanation. Method Two uses simple equations to compute the expected value of the options. In a sense, Method Two is more straightforward because you just use numbers. However, some students may not be able to do the math, or they may prefer a more abstract way of evaluating the information. In these cases, Method One is more appropriate.

compare goals in the abstract. For example, don't debate whether the goal of saving human lives is more important than saving trees. Would anyone choose to save a tree rather than a person? But what if the choice were between saving all the forests in the eastern United States compared to saving ten people from a fatal disease? It's important to compare specific advantages such as this rather than general goals.

Importance bars

Compare advantages on different goals by using the length of a line. Have students draw lines of varying lengths to indicate their individual judgments about an advantage's importance. This is just one of several ways to put values on a scale. Line lengths can be translated into numbers, or the numbers can be estimated directly. Specifically, students should draw a long line (importance bar) representing the advantage that seems most important across all goals. Then, students draw shorter lines for the other advantages.

For example, suppose a student thinks of only three goals and two options, as shown in Chart 4A (page 17). The numbers are only hypothetical examples. The student should draw importance bars to show the advantage of the best option (+) over the worst option (-) for each goal. These bars are simply the differences between the best and worst options in the figure. They are the student's subjective ideas of how important each advantage is compared to the other two.

In Chart 4B, the student thought saving two percent of U.S. trees was the most important advantage and therefore drew the longest line for "Save trees." The student rated saving 130 human lives per year almost as important as saving two percent of U.S. trees. Creating 800 more jobs was rated only less than half as important. The bars show that the combined advantages for Option B (two bars laid end to end) are a little greater than the advantage for Option A.

If the student doesn't agree with this conclusion, then he or she should reevaluate and redraw the lines.

Importance bars can be used for any number of goals and options. The importance bar for a goal should have the worst option on the left end and the best option on the right end. The procedure for drawing importance bars is described below.

Procedure

(A) Suppose the decision chart has three options and four goals, as shown in Chart 5. First, students mark the best option on each goal (+) and the worst option (-).

(B) Students compare advantages across goals using only the best and worst options. How important would it be to have a (+) rather than a (-) on each goal? How important is the *advantage* of the best option over the worst option on a given goal? Suppose students decide that the biggest advantage is on Goal 3 (B vs. C). They draw a long line to the right of Goal 3, placing C at the "Worst" end and B at the "Best" end.

They repeat the procedure for each of the other goals. They only compare the best and worst outcomes for each goal, and for now ignore any other options. (Which options are best and worst will change for each goal, as illustrated in Chart 5.)

For goals other than Goal 3, the advantages of (+) over (-) are less important, so they will be shorter lines. Have students think carefully about the length of each line. The length should indicate the importance they assign to that advantage. The lengths of the shorter lines should make sense when added together. For example, suppose a group decided the Goal 3 advantage is most important, but the advantages on Goals 1 and 2 together are more important than the Goal 3 advantage. This implies that if the Goals 1 and 2 lines are laid end to end, the result is longer than

the Goal 3 line. How *much* longer should indicate how much more important those combined advantages are than the Goal 3 advantage.

(C) Compare the other options on each goal (if there are more than two options). Since there are three options in this example, there will be an "in-between" option. Is it closer to the best, or to the worst? Suppose a group studies the Goal 1 outcome for Option B and judges that it is just a little closer to Option A (worst) than to Option C (best). So, they place B just a little to the left of the center of that line, as shown in Chart 5.

(D) Complete the analysis by adding up the lengths of the importance bars. The best option is the one with the longest combined bar. First, compare just two options at a time, such as A and B. Measure the lengths of the line segments A—B and B—A on each line. Add all the segments favoring A (B—A), then all the segments favoring B (A—B). Is the A sum greater than the B sum? If so, A is better according to this analysis.

Before students compute their results, have them check to see if their line lengths are right by asking them questions such as, "Your lines for Goals 3 and 4 are each almost as long as your line for Goal 2. Do you really think that the advantage of Best over Worst on Goals 3 and 4 added together is more important than the advantage on Goal 2?"

If students disagree with what the bars imply, the error may be in how the bars were drawn. Should their lengths be reconsidered? Do they accurately reflect the students' feelings about the advantages of each outcome?

Have students practice making tradeoffs using their importance bars to compare advantages. This can be done graphically (without numbers). Have them lay a blank sheet of paper beside the decision chart and draw importance bars with a pencil. When the students have finished their

Importance Bars

DECISION CHART 4A: Comparing Outcomes

	Option A	Option B
Save trees	*1% of U.S. trees die (+)*	*3% of U.S. trees die (-)*
Keep jobs	*200 more jobs created (-)*	*1,000 more jobs created (+)*
Prevent human deaths	*50 more deaths per year (-)*	*Save 80 lives per year (+)*

DECISION CHART 4B: Drawing Importance Bars

	Advantage	Importance Bar
		Worst Best
Save trees	*Save 2% of U.S. trees*	*Option B—————————————Option A*
Keep jobs	*800 more jobs*	*Option A—————Option B*
Prevent human deaths	*Save 130 lives per year*	*Option A———————————Option B*

DECISION CHART 5: Drawing Importance Bars for Three Options

	Options			Importance		
	A	B	C	Worst		Best
Goal 1	-		+	A B C x x x x x x x x x x x x x x x		
Goal 2	+		-	C B A x x x x x x x x x x x x x x x x x x x x x x x x x		
Goal 3		+	-	C A B x x x x x x x x x x x x x x x x x x x x x x x x x x x x x x x x x		
Goal 4		-	+	B A C x x x x x x x x x		

bars, have them cut the paper into strips, each strip the length of a bar (with option labels showing). Have the students lay all the bars favoring Option A end to end. Do the same for the other options. Compare the combined lengths. The preferred option is the one with the longest combined strip.

Fast way to get total

There is a faster way to compare segments for two options at a time. Make the left (worst) end of every line equal to zero. Then, to evaluate any given option, simply measure the distance from the left end to that option, and add the numbers for all goals. The best option will have the highest number.

On Chart 5, we count the number of x's (measuring the line lengths with a ruler would be faster). The table below gives the totals.

	Options		
	A	B	C
Goal 1	0	7	15
Goal 2	26	7	0
Goal 3	25	32	0
Goal 4	4	0	9
TOTAL # of x's	55	46	24

From the totals, it appears that A is slightly better than B overall, and that C is worst. Note that some shortcut methods will give misleading results if they don't take into account the importance of each advantage. Example: Choosing the option with the one biggest advantage would lead you to choose B. Second example: Choosing the option rated best on the most goals would lead you to choose C.

Incorporating probability

After students are familiar with importance bars, they can take probability into account. If the probability is expressed on a continuum, have them use the most likely outcome and ignore the range of other probabilities. Then, have them draw an importance bar based on the most likely outcome. If the outcome box is expressed as an all-or-nothing event, have students estimate the two values (yes and no) separately by drawing two importance bars and then picking a value in between. The in-between value should reflect the probability of each outcome. A long bar and a short bar (two bars for one outcome) should be "averaged," and the new bar length should be closer in length to the more likely outcome.

Analyzing the Decision: Method Two

In Method One, importance bars helped us assign values to each outcome. Method Two uses the importance bar values and probability estimates to compute the total expected value for an option.

To obtain a numeric value for each outcome, use the importance bars and convert them to numbers (see "Fast way to get total"). Another way to get numbers is to ask students directly. Have students assign the number zero to the worst outcome on each goal and compare advantages across goals. However, numbers should be used with caution. For example, if you ask students to put all values on a scale of one to ten, they cannot express the value of an advantage that is 100 times as important as another.

For each outcome box, enter the probability (P) and value (V) data. If an outcome box has more than one possible outcome (all-or-nothing events), put separate value and probability numbers for each. For continuum events, use the most likely outcome as though it were certain (P = 100 percent). Multiply the P and V data for each outcome box and add up all outcomes for the option (column). This is the total expected value for the option. The final decision should be the option with the greatest expected value.

Checking Your Work: Sensitivity Analyses

A sensitivity analysis asks the question: "How sensitive is my decision to uncertainty about the outcomes?" For example, would the same option still be the best if you assume that the outcome least favorable to it occurs?

If students identify only one goal for which sensitivity analysis swings the decision, they may want to research that goal further. Then, they may reach a well-grounded recommendation for action.

Procedure

(A) Recall which option scored highest when students combined all the importance bars or computed expected value. Let's call it Option W for "Winner." Circle option W in green.

To check this option, lay a clean piece of paper beside your chart. Try changing just one uncertain outcome to the least favorable end of the range. If you have an all-or-nothing event, assume the outcome least favorable to W. If you have a confidence interval, choose the end of the interval least favorable to W. Circle or label these new outcomes in red to show that these are "least favorable" assumptions.

(B) Redraw the importance bars or recompute expected value. Use the "least favorable" values for this goal, but keep the original values (or importance bars) for the other goals. Calculate the total merit of each option. Does Option W stay the best? If an outcome by itself can change Option W from winner to second place (or worse), just by changing the outcome from the initial estimate to least favorable, write "Needs research" in the box. Concentrate research on this outcome.

Final Decision

Using either importance bars or expected value, students pick an option for the decision. If you feel your students are not ready to handle sensitivity analyses, then end the decision-making process with the importance bars or expected value result.

To set priorities about which research would make the decision more conclusive, ask each group to pick the most uncertain outcome or box of the chart. Then, have them analyze uncertainty only in that box. Explain that a more complete analysis would look at all uncertainties in the chart.

When analysis and gut feelings don't agree

If students feel the decision is not the one they would really make, ask them if an important goal has been left out of the analysis. Do they think a fact or estimate is wrong? If such a factor can be identified, ask them to make a new estimate or value judgment, or add a new goal, and do the analysis again.

There may not be a logical answer when analysis and feelings don't agree. Sometimes people waver back and forth between preferences, depending on their mood or which consequences are on their minds at the moment. One of the purposes of Decision Analysis is to get people to face up to such inconsistencies and try to decide which decision would best satisfy their goals in the long run.

FINAL DECISION ESSAY

Have students write a short essay on the final decision. Did it agree with their pre-analysis prediction? Why or why not? What did students personally learn from the decision?

Step Four Summary

Key concepts

♦ Analyze the decision using the goals, options, outcomes, and probability.

♦ *Importance bars* provide a visual representation of the *advantages* of each option for each goal. The preferred option is the one with the greatest combined advantage for all goals.

♦ *Sensitivity analysis* allows decision makers to determine if a particular uncertainty affects the decision as a whole.

♦ Gut feelings and science-based decisions may not agree. In these cases, some considerations may be missing.

Procedure

Decision chart analysis

(A) Method One: Draw importance bars based on the advantages of options for each goal. Add the importance bars. Which option is the winner?

(B) Method Two: Assign numerical values and probabilities for each outcome in the chart. Compute expected value. Which option is the winner?

(C) Sensitivity analysis: Check the uncertainty of outcomes to identify research needed.

Final decision: Any additional issues?

(A) Discuss the process. Are students happy with the final decision? Why or why not? Were there missing factors?

(B) What research is needed to make the decision more conclusive? Identify the single most important fact.

Notes

1. National Research Council. 1996. *Understanding Risk: Informing Decisions in a Democratic Society.* Paul C. Stern and Harvey V. Fineberg, eds. (Washington, DC: National Academy Press), 215.

2. For a full discussion of effects of hazards, see *Understanding Risk,* 44–49.

3. The three ethical concepts come from *Understanding Risk,* 40–42.

4. William Freudenburg. 1996. "Risky Thinking: Irrational Fears About Risk and Society." *The Annals of the American Academy of Political and Social Science: Challenges in Risk Assessment and Risk Management* 545 (May), 46.

TEACHING PLAN:
Part One

Objective

This teaching plan is designed to provide an overview of decision making for you and your class to use for all the activities in this book, or for decisions encountered in any science topic you cover. You may expand upon any of these steps when you conduct Part Two activities.

Time Management

An introduction to decision making can be completed in two to three 50-minute sessions.

Procedure

First session

1. Introduce the topic of decision making with the activity, "Planning the Big School Event, Part One" (pages 21–24). (20 minutes)

2. Conduct a general class discussion on decision making. Hand out "Summary of Decision Making" (page 119). Students should follow along as each topic is discussed. (25 minutes)

Some important topics are:

♦ Decisions may be evaluated from different perspectives. See teaching box, page 6.
♦ Risk consists of the *exposure* and the *intensity of effect* of a hazard to a given population, material, or system. See teaching box, page 6.
♦ Common *effects* of hazards include: human health, ecological, economic, government, and ethical. A special kind of effect is *risk space*, which is the public's reaction to a hazard based on its characteristics (controllability and familiarity). See page 7.
♦ A *goal* is a criterion of value which helps decision makers evaluate options. See teaching box, page 9.

3. Homework. Explain the concept of a hazard (page 6). Have students make lists of hazards. Note: This list is part of the activity, "Public Perception of Hazards" (pages 25–27). You will complete the activity in the second session. (5 minutes to explain)

4. Have students keep a journal of their work. Guidelines for keeping a journal are found on page xi. This journal will be used in all the following activities for note taking, and the journals will provide you with a method of student assessment.

Second session

1. Conduct student activity, "Public Perception of Hazards." Use the lists the students made for homework to complete the grid. (20 minutes)

2. Conduct student activity, "Planning the Big School Event, Part Two" (pages 28–31). (30 minutes)

Third session

1. Discuss the difference between all-or-nothing and continuum probabilities. See pages 13–14. (10 minutes)

2. Discuss how to analyze a decision using importance bars and expected value. Use page 17 as an overhead master or handout. (20 minutes)

3. Discuss with students some basic research techniques, such as Internet searches, encyclopedias, periodical searches, and basic library skills. (20 minutes)

Extension

The following topics are suitable for advanced students.

♦ Extension idea in the activity, "Public Perception of Hazards" (pages 25–27)
♦ Advanced probability steps (page 14)
♦ Sensitivity analysis (page 18)

Assessment

Have students write a short summary of decision making. Students could use the Summary of Decision Making (page 119) as a guide. Have them write about a time when they made an important decision. How did they decide? Was it a similar method to their summary of decision making? Why or why not?

Resources

See Selected Resources (page 127) for more information on decision making.

TEACHER PREPARATION

Review Part One; note teaching tips in margin boxes.

Planning a Big School Event, Part One

Student Handouts

Student section (pages 23–24)
Students will also need to keep journals. See page xi for instructions.

Objectives

Students will:

♦ Discuss the way they already make decisions.
♦ Learn to consider value and probability when weighing outcomes in decisions.
♦ Learn that decision making is enhanced by deliberation and consultation with others.

Procedure

See Teaching Plan: Part One (page 20).

Part A

1. Have students construct their decision-making journals. During the activity, let students experiment with brainstorming and note taking. Then, give them feedback on ways to improve.

2. The situation:

Event: A big school event, such as a charity festival or graduation ceremonies.

Options: Hold the event outdoors, or hold it indoors in the cafeteria.

Goal: Maximize enjoyment of the festival.

Outcome: *How much* the people who go to the festival enjoy it. Whether or not it rains will greatly affect the outcome if it is held outdoors.

Describe the big school event to students. Younger students could discuss a spring or fall festival, or a charity event with games and food booths. Planning a graduation ceremony might be more appropriate for older students. Let the students quickly list some sample entertainment that will be at the event. Make sure the entertainment could be held either indoors or outdoors. Some examples could include: Halloween games (bobbing for apples, etc.), games involving running and/or throwing bean bags, circle or line dancing, couple

dancing, or eating. Students may add any additional details, such as hours or how many people attending, that will help them imagine the event.

Don't get bogged down in planning the event. This is just an example to practice making decisions.

3. Have students read "The Big School Event" in the student section. You may wish to adapt the scenario as necessary for a graduation ceremony or other events.

4. Class Discussion: Ask several students to relate to the class how they decided. Use their answers to demonstrate how they took into account value and probability (see below). Explain that weighing outcomes by estimating their probability and value is the basis of decision making.

VALUE

Some students are likely to mention how good or bad outcomes are in their own words (*e.g.*, "Rain on the party would be terrible, so..."). Students will easily see that the outcomes can be ranked from best to worst as follows:

Best: The festival is held outdoors and the weather is good.
Next best: The festival is held indoors.
Worst: The festival is held outdoors and it rains.

These three outcomes are ranked by their "value." Other words for value are "preference" and "utility." Economists like to express value in terms of how much money an outcome is worth. One problem with doing this is that people value money differently; another is that people often find it distasteful to equate serious human events (like saving a life) with a dollar value.

PROBABILITY

The uncertainty about rain is a key factor in the decision. Have students read "The Probability of Rain" in the student section. This explains how the same numerical probability can be called different things. Stress that numbers are used to eliminate interpretation from probability estimates.

If one person says rain is "likely" and another agrees, ask them to assign a number percentage to that estimate (*i.e.*, 20 percent chance of rain). Discuss what a "20 percent chance of rain today" means. Explain that it means that, based on meteorological data, it will rain about one day in five on "days like this."

Part B

1. Estimate outcomes: Have students write down on the student handout what they think the percent chance of rain is on one day in the month, the day when the big school event is planned. Then, they should decide Outdoors or Indoors. For more realism, students could consult rainfall records for your area. Record the class answers on the board or on an overhead. Ask students if there is any relationship between their decisions and the probability estimates. Decisions to plan it outdoors may be associated with lower probabilities of rain. Have some students explain why they decided the way they did.

Once students are clear on the two concepts of *value* and *probability*, show them how some students related the two ideas in reaching a decision. For example, a student might say, "Rain would ruin the party, but it is so unlikely that I would take a chance and plan it outdoors." Or, conversely, "Rain is not very likely, but if it did rain it would be such a disaster that I would plan the festival indoors."

2. Conduct a brief discussion with students about other decisions they have made.

♦ What was each student's first decision of the day? Examples might be: which clothes to wear, whether to make the bed or not (inaction is actually a decision not to act), what to eat for breakfast, and other such morning activities.

♦ How important are their decisions? Compare student decisions to those made by the mayor of your town or the President of the United States.

♦ Have students discuss why having many people involved in decision making can be an advantage. Could it be a disadvantage? Point out that few people ever fully agree; the decision may not get made if people cannot compromise.

Assessment

Assign "How Do You Make Decisions?" for homework. Keep a list of any techniques students mention for comparing options. Later you may use these examples to connect the methods in this book to how students already think. Later (page 28), we will pick up this activity again to expand on these concepts.

Planning A Big School Event, Part One

The Big School Event

Suppose we are going to have a big school event. You are in charge of planning it, and the main decision you have to make is whether to have it outdoors or indoors. The school has a beautiful area of grass and tress that would be perfect for the event, but if it rains everyone will wish it had been indoors. The cafeteria is adequate for a party, but not as good as outdoors if the weather is pleasant.

The games and contests that students want to hold cannot easily be moved indoors if it rains. You cannot schedule a raindate, because the school schedule is overbooked. So, you must decide three weeks in advance whether to hold it indoors or out. Weather cannot be forecast that far ahead with complete accuracy.

What should we do? It's up to you to decide. The decision is: Do we hold the event outdoors or indoors? In your journal, describe how you might reach a decision.

The Probability of Rain

Will it rain the day of the school event? What are the chances? The probability of an event can vary from "Impossible" to "Certain to happen." Probabilities are useful in a variety of situations. A person deciding whether to run for public office will consider the probability of winning as a

Numerical	Percent chance	Description
1.0	100%	Certain to happen
0.9	90%	Very likely
0.8	80%	
0.7	70%	Likely
0.6	60%	
0.5	50%	50-50 chance
0.4	40%	
0.3	30%	Unlikely
0.2	20%	
0.1	10%	Very unlikely
0.0	0%	Impossible

major deciding factor. Without using numbers, a coach or player weighs the probabilities that different strategies will be successful before choosing one.

A numerical scale from zero to one is often used to express probability, where zero means "impossible" and one means "certain to happen." Another scale often used in reporting weather is the percent chance scale, where zero percent means impossible, and 100 percent means certain to happen. A "50 percent chance of rain" means that rain and no rain are equally likely.

There are words at several places on the scale. But except for "certain" and "impossible," even experts can't agree on which words should correspond with each number on the scale. For example, some might say that 20 percent is "unlikely," and some might say that 20 percent is "very unlikely."

What words should be used to describe very rare events that have only one chance in a million of happening (probability = .000001)? Numbers allow finer distinctions than words among probabilities. For example, an event with a probability of one in one hundred (.01) is *ten thousand times* as likely as an event with a probability of .000001. Yet both of these events might be described by the same words, such as "extremely unlikely."

Number scales are used so people can communicate more clearly the likelihood of an event. Also, numbers can be used in mathematical formulas to calculate probabilities in complex situations.

The enjoyment of the event you are planning will depend on its location. But another consideration is: will people enjoy the festival if it rains? What is the probability of rain?

What do you predict is the percent chance of rain? Write it down here: _____%
Now, circle your decision: Hold festival **Outdoors** or **Indoors**.

How Do You Make Decisions?

Write short paragraphs in response to the questions below.
   (A) What are the most important steps in making a decision?
   (B) How would you make a "really important" decision?
      Select such a decision and list the factors you would consider.

Public Perception of Hazards

Student Handouts

Student section (page 27)

Objectives

Students will:

♦ Learn that all hazards are not equal in the public's mind, even if those hazards have the same risk.

♦ See how students react differently to similar risks with different characteristics.

♦ Learn that control over a hazard, and if it is observable and understood, are important considerations for determining which hazards are studied and reported.

Procedure

1. Hand out page 27. As a homework exercise, students should brainstorm in their journals a list of 20 hazards they deal with directly or indirectly in their lives. Examples of hazards could include: riding a bicycle, flying in a plane, eating fatty food, pollution, pesticides on food, and papercuts. Encourage them to think of events they have heard about through the news or from friends and family.

2. As they make their list, have them rate each hazard in order of how frightening or dreadful each is, most dreadful on top, least at bottom. Stress to students that it doesn't matter how dangerous the hazards are in reality or mathematically, just rate them in order according to how the students feel about the hazards.

3. Either in groups or together as a class, have students use the criteria to place their hazards into the four-quadrant grid. Most probably, their hazards will fall into place much like the chart on page 8. You may want to start them on the process by using an overhead of the blank chart (page 27) with one or two examples.

4. Class discussion. Ask students the following questions.

What connection do you observe between the hazards that you listed as most dangerous and the quadrant in which you placed the hazard? Is there a pattern? Students will generally find that the most dreadful hazards are in the upper right quadrant. These are hazards that are uncontrollable, not well understood, and affect many people. They are more frightening than everyday, controllable hazards.

What does the grid tell you about how certain hazards are seen by the public? Some hazards cause a more emotional response than others depending on where they fit into the risk space grid. These are the ones for which we usually desire government regulation.

How would risk space affect your decision making? Answers may vary. Ensure that students see how doing a risk assessment is important when evaluating a hazard. Have them consider how preconceived notions they have may be wrong.

If students' hazards do not fit quite like the risk space grid on page 8, use some of the student examples to explore why. For example, a student may have some special knowledge which allows him or her to have a different opinion about a particular hazard. Or perhaps a student misunderstood the nature of the hazard.

You may modify this activity by including only hazards which pertain to your field of study. Students can then use information they learned in class to brainstorm hazards. For example, biology hazards could be genetic mutations, the effects of pollution on animals, nutritional deficiencies, and viruses.

> **NOTE**
> *Before doing this activity, review "Public Perception of Hazards" (page 9).*

Extension

Students may research the actual risk of a hazard and compare the perceived risk to the actual risk. Some easy-to-use books are listed below in Resources.

Assessment

Use the Extension for assessment of student work.

Resources

Laudan, Larry. 1994. *The Book of Risks: Fascinating Facts About the Chances We Take Everyday*. New York: John Wiley & Sons, Inc.

Siskin, Bernard, Staller, Jerome, and Rorvik, David. 1989. *What Are the Chances? Risks, Odds & Likelihood in Everyday Life*. New York: Crown Publishers, Inc.

Skolnick, Susan A. 1985. *Book of Risks*. Bethesda, MD: National Press, Inc.

For teachers and college students:

Freudenburg, William R. 1996. "Risky Thinking: Irrational Fears About Risk and Society." *The Annals of the American Academy of Political and Social Science: Challenges in Risk Assessment and Risk Management* 545. May: 44–53.

Morgan, M. Granger. 1993. "Risk Analysis and Management." *Scientific American*. July: 32–41.

Slovic, Paul. 1987. "Public Perception of Risk." *Science* 263:288–285.

Public Perception of Hazards

Procedure

1. Brainstorm in your journal a list of 20 hazards you have heard about or experienced. As you make your list, rate each hazard in order of how frightening or dreadful each is, most dreadful on top, least at bottom.

2. Use the criteria to place the hazards into the four-quadrant grid below.

Questions

◆ What connection do you observe between the hazards that you listed as most dangerous and the quadrant in which you placed the hazard? Is there a pattern?
◆ What does the grid tell you about how certain hazards are seen by the public?
◆ How would risk space affect your decision making?

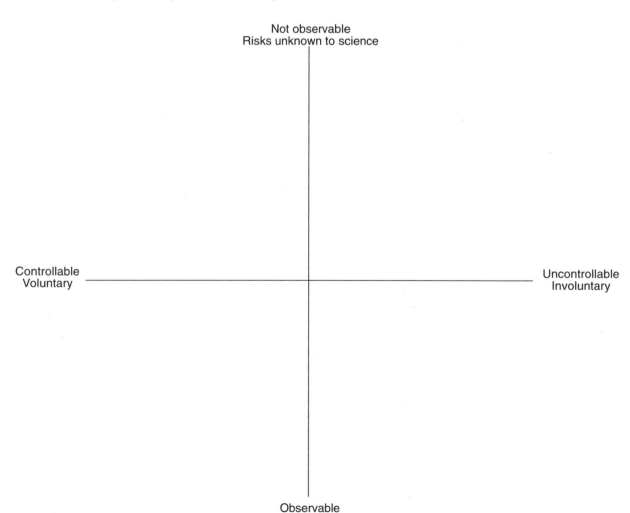

Planning a Big School Event, Part Two

Student Handouts

Student section (pages 30–31)

Objectives

Students will:

♦ Make decision charts.
♦ Learn about expected value and apply their learning to the big school event example.

Procedure

1. Decision charts. Remind students of the big school event they planned in Part One (pages 21–24). The options for the event are outdoors or indoors. The outcomes are enjoyment of the festival depending on if it rains or not. See decision chart in student section.

   Remind students that the outcome box for outdoors has two factors because there is uncertainty (will it rain or not?). The total expected value for the option will be the expected values of both factors added together. Expected value is explained on page 12 and below.

2. Expected Value. Now we have a decision chart for the big school event. But how do we compare the different outcomes? How do we rank or grade them according to how important they are?

   Use the reading in the student handout on expected value to discuss the concept with your students. Stress that this is a way to make a decision using mathematical principles, but that they should always check what they believe should happen against what is happening according to the computation. If there is a discrepancy, the student may have left out a consideration from the model.

3. Assigning value to outcomes. Students should use Figure 1 in the student handout to assign the value for "Indoors." How high or low on the scale they check should reflect their feelings about how much people will enjoy the event if it is held indoors, compared to the two possible outcomes if it is held outdoors. Remind students that it doesn't matter where on the real number line they put the best

and worst outcomes, as long as each unit on the scale stands for the same size increase in value.

Tell students that if they think the cafeteria is a terrible place for a big school event, almost as bad as being rained on outdoors, they should place "Indoors" very near the bottom of the scale. If they think the cafeteria might be a neat place for the event, perhaps because it has a good dance floor, then they might place "Indoors" very near the top.

4. Class discussion. Have students answer Question 2 on the handout.

   Question 2. Suppose you know that there is a 50 percent chance of rain on the day of the big school event.

   (A) Knowing this, would you choose outdoors or indoors for the event? Two reasonable answers: Indoors, because the setting doesn't matter as much as having fun and eating. Outdoors, because the whole idea of an event is the outdoors, so I would take a chance on the rain.

   (B) Does your choice relate to where you placed "Indoors" on the value scale? In what way? Reasonable answer: Yes, the higher Indoors is, the less difference there is between Indoors and Outdoors, and the more likely you are to choose Indoors where the rain can't spoil the party.

5. Computing expected value. Students should now apply expected value to the big school event using the decision chart. Use the example in the student handout (page 31).

   Answer: Indoors: EV = 2. Rain doesn't affect the indoors party, so we don't compute probability.

   Outdoors: EV = (P of rain × 0) + (P of *no* rain × 20); EV = (.50 × 0) + (.50 × 20) = 10

   So, the EV for Outdoors is 10 and the EV for Indoors is only 2. It makes sense that the expected value of having the event Outdoors is higher, since the student doesn't like the cafeteria.

Assessment

Assign these questions for homework.

1. The probability of rain is 50 percent. Flora, who believes the cafeteria is a nice place for the event, gives a value of 18 to the Indoors option. (A) Calculate the expected values for Indoor and Outdoor events. (B) Which option would Flora choose if she uses expected value to decide?

   Answers: (A) Indoors: EV = 18; Outdoors: EV = 10; (B) Indoors, because 18 is greater than 10.

2. Maynard, another student, rates Indoors a 9 in value. (A) Which option do you think he would choose? (B) Which option do you think Maynard would choose if the probability of rain were 70 percent?

Answers:

(A) Outdoors, if he uses expected value to make his decision.

(B) Indoors: EV = 9

Outdoors: EV = (.70 × 0) + (.30 × 20) = 6

He would choose Indoors if he uses expected value (9 is greater than 6).

3. If the chance of rain is 50 percent, where would you put Indoors on the value scale to make the decision difficult? Why?

   Answer: Put Indoors at 10, halfway between the best and worst outcomes. This makes the EV equal for the two options.

Planning a Big School Event, Part Two

DECISION CHART:
Big School Event

	Option: Hold event indoors	Option: Hold event outdoors
Goal: Enjoy event	*EV (Indoors enjoyment) =* *V (enjoyment inside) × P*	*V (enjoyment, rain) × P (chance of rain)* *V (enjoyment, no rain) × P (chance of no rain)*
	Note: P = 1, because rain doesn't affect the event	*Note: add top and bottom lines to obtain EV*

Expected Value

We say that two options have the same *expected value* if the choices are likely to lead to equally good results. The expected value (EV) of an outcome is the probability (P) it will happen multiplied by the value (V) of the outcome. The equation is: $EV = P \times V$.

If there are two possible outcomes to a choice, the value of each outcome is weighted by its probability. This can be demonstrated with money. Suppose a relative offers to let you flip a coin. If it comes up heads, you win $1.00. If it comes up tails, you win nothing.

The probability of the coin coming up heads is 0.5. The same is true for tails. (Half the time it will come up heads and half the time tails.) The expected value of playing the game is 50 cents, as shown below.

Expected value of game = $(P \times V)$ for heads + $(P \times V)$ for tails

$EV = (.5 \times \$1.00) + (.5 \times \$0.00) = 50$ cents

In other words, the expected value is what you would expect to win on the average in each game if you played the game many times. About half the time you would win $1.00, and about half the time you would win nothing. The average amount you win would therefore be around 50 cents. Another example of finding expected value is shown above in Decision Chart: Big School Event.

Assigning value to outcomes

Obviously, to compute expected value, you must give a number grade to an outcome. In this example, give the worst outcome a value of zero, and the best outcome a value of 20.

1. What should the value be for Indoors? Write it in Figure 1.

2. Suppose you know that there is a 50 percent chance of rain on the day of the big school event.

   (A) Knowing this, would you choose outdoors or indoors for the event?

   (B) Does your choice relate to where you placed "Indoors" on the value scale? In what way?

FIGURE 1:
Assign Value

Value	
20	Outdoors, good weather
19	
18	
17	
16	
15	
14	
13	
12	
11	
10	
9	
8	
7	
6	
5	
4	
3	
2	Example value for Indoors
1	
0	Outdoors, and it rains

Computing expected value

Example: Suppose a student thinks that the cafeteria is not a good place for a festival, so the value of Indoors is only a "two." Now compare the expected value of Indoors and Outdoors, the two options in the decision. Assume the probability it will rain is 50 percent.

	Option: Hold it indoors	Option: Hold it outdoors	
Goal: Enjoy the event	Enjoyment (V) = 2 P = 100%	No Rain: V = 20, P = .50	Rain: V = 0, P = .50

Calculate:

EV Outdoors = _____

EV Indoors = _____

Questions

1. The probability of rain is 50 percent. Flora, who believes the cafeteria is a nice place for the event, gives a value of 18 to the Indoors option.

   (A) Calculate the expected values for Indoor and Outdoor events.

   (B) Which option would Flora choose if she uses expected value to decide?

2. Maynard, another student, rates Indoors a 9 in value.

   (A) Which option do you think he would choose?

   (B) Which option do you think Maynard would choose if the probability of rain were 70 percent?

3. If the chance of rain is 50 percent, where would you put Indoors on the value scale to make the decision difficult? Why?

PART TWO

Guided

Activities

TEACHING PLAN: Guided Activities

Objective

Each Guided Activity provides a step-by-step case study in decision making. They are not intended as full scientific discussions of each topic, but instead summaries of the problem, with room for research and expansion as desired. There are no correct answers to the activities. This may cause some concern to students who are used to coming up with a correct answer. Stress that, for some of these topics, science and society have not found an answer, either.

Required Student Skills

Students should be familiar with the Part One concepts. See Teaching Plan: Part One (page 20).

Time Management

The activities usually can be completed in two 50-minute sessions with a few out-of-class days of research and thinking assigned as homework. Times given in the General Procedure can be modified to fit the needs of your class. Use the General Procedure as a guide. Some activity procedures are slightly modified from this format. You may want to conduct the second session several days after the first session to give students time for research.

General Procedure

First session

1. Before beginning a new activity, photocopy and hand out three items: the student section for the activity, Appendix B: Summary of Decision Making, and Appendix C: Decision Chart.

2. Have all students read the handout. (5 minutes)

3. Conduct a brief class discussion of the topic under consideration. Make sure students understand the issue. (5 minutes)

4. Have students write brief, preliminary decisions in their journals. Encourage them to use what they know of the decision-making process. (5 minutes)

5. Continue the class discussion on the issue. (40 minutes)
   - Ask the question: Whose decision is it? See teaching box (page 6).
   - Research and discuss a risk assessment for the hazard. Note: An additional day could be spent on research and discussion to fully define the hazard. See pages 6–9.
   - Make a preliminary list of goals for the decision. See teaching box (page 9).
   - List some possible options (these need not be final). See teaching box (page 11).
   - Elicit ideas about what kinds of data and additional information students might need to make a more informed decision. You may want to keep a running list of student ideas on the chalkboard.

6. For homework, have students conduct research using the ideas generated in the class discussion to guide them. They should create a decision chart with the appropriate goals, options, and outcomes. If students are familiar with probability, have them make probability estimates for the outcomes.

Classroom management: Students interested in particular research areas may form groups on their own, or you may wish to divide students into groups and assign areas for them to research. Research could take several days of out-of-class work.

Second session

1. Conduct a classroom discussion on the research results. What facts turned out to be most pertinent to the decision? What parts of the decision are value-based? (10 minutes)

2. Divide students into groups and have them use their decision charts to weigh the options using importance bars or expected value. Have students decide as a group. (25 minutes)

3. Lead a class discussion of the group decisions. (15 minutes) The discussion should include questions such as:

   ◆ What option did they pick? How does the final decision compare to the preliminary decision? Are they different? Why or why not?

   ◆ What uncertainties in the outcomes make the decision difficult?

   ◆ What single most important fact is needed to make the decision more definite?

Extension

Extensions are related ideas for another decision, or to get students to think more deeply about the issue. Extensions can be assigned as assessment activities. Students may have their own ideas about extending the activity, which may be developed in class or as homework as time allows.

Assessment

Student journal entries may be used for assessment. You may also assign the Extension. See Appendix D: Assessment Rubrics.

Resources

Suggested books and Internet sites for teacher preparation or further reading on the topic are provided in each activity.

Xenotransplants

Background

Perhaps you have read *Frankenstein* or have seen one of the movie versions. The story is about a dead human brought to life by transplanting organs from other humans. When Mary Shelly wrote the book in 1818, the tale of *Frankenstein* seemed far-fetched, but today doctors routinely save lives using organ transplants.

Over 30 years ago, the first human heart transplant was successfully performed. The heart was transplanted from a donor to a recipient in South Africa. Since then, doctors worldwide have conducted thousands of heart transplants. Kidneys, lungs, and livers can also be transplanted.

Each year in the United States, approximately 1,000 people receive donor hearts. The success rate for transplants is relatively good, but waiting lists remain long due to the lack of human donor organs. There is a waiting list of more than 40,000 patients for heart transplants, and 24,000 are on wait lists for kidney transplants. Nearly 3,000 people will die each year waiting for an organ.

In an effort to increase the supply of organs, researchers have experimented with xenotransplantation, the process of transplanting organs between different species. Experimenters have placed baboon hearts into humans, but the results have been poor so far. For the past 15 years, parts of organs have been transplanted from pigs to humans. For example, pig heart valves can be successfully transplanted into humans. Pig livers have been temporarily connected to human beings outside the body to purify blood, with good results.

The main risk of xenotransplants is a phenomenon called hyper acute rejection (HAR). This occurs when the human recipient develops antibodies that reject the organ. The organ then becomes engulfed in blood clots and ceases to function.

One solution may be to use genetically altered pigs. The pigs are given a human gene that makes the pig organ seem more human to the immune system. The gene is called decay accelerating factor (DAF) which prevents rejection by the body. Hearts from pigs that have not been genetically altered survive an average of 55 minutes when transplanted in monkeys. Genetically altered hearts survive an average of 40 days. Two hearts have functioned for more than 60 days. These results have given considerable hope to researchers who wish to overcome the obstacles of xenotransplantation. But whole pig hearts have never been transplanted into humans.

Another risk of xenotransplants is the chance that unidentified animal viruses could be transferred to humans. The virus could spread from the human recipient to other people, potentially causing an epidemic.

Organ transplants are expensive, requiring extensive medical care for recipients even after a successful operation. Xenotransplants are even more expensive because of the experimental nature of the surgery and the drugs needed to counteract rejection and infection. Some people believe that this money would be better spent learning how to prevent the diseases that cause the organs to fail. For example, some researchers believe that most heart disease can be prevented through lifestyle changes, such as exercise and better diet.

Some people have moral objections to genetically engineering pigs with human genes. They question how much scientists should combine human and animal genes. Others think that using animals as "organ farms" is wrong because the animals suffer.

The Decision

The U.S. Food and Drug Administration (FDA) is reviewing its policy on xenotransplants. Even though pig valves have been transferred for several years, some virologists have said that there may be a risk of transferring pig viruses to humans. Suppose you are a member of the FDA committee studying this issue. You must decide whether or not to allow experimental transplants of genetically altered pig hearts or valves into humans. One option could be just to allow pig valves to be transferred.

TEACHER SECTION

Xenotransplants

Student Handouts

Student section (pages 36–37)
Appendix B: Summary of Decision Making
Appendix C: Decision Chart

Objectives

Students will:
- ♦ Learn what xenotransplants are and conduct a risk assessment using the student handout.
- ♦ Make a decision about experimenting with xenotransplants to humans.

Terms for Understanding

Xenotransplant, antibody, immune response, rejection, genetic engineering

Procedure

Part A

1. Have students read "Xenotransplants" (student section). After a brief discussion of the topic, students should write a preliminary decision in their journals.

2. Discuss The Decision (in student section). Whose decision is it? The students are to assume the role of a member of a U.S. Food and Drug Administration (FDA) committee which will decide whether or not to allow experimental transplants of genetically altered pig hearts and/or valves to humans. Since current research focuses on transferring pig hearts to baboons, this decision anticipates when experimenters will want to try the procedures on humans.

3. Risk assessment. Discuss how the public sees this issue based on risk space (not well understood, new technology, some ethics involved). Have students brainstorm some of the major issues. Some examples are:

- ♦ Lack of human donor organs and long waiting lists.
- ♦ Chances of rejection of donor organs.
- ♦ Chances of introducing animal diseases into humans.
- ♦ Quality of research (relatively unsuccessful clinical trials).
- ♦ Quality of life of people who have received transplantation.
- ♦ Moral aspects of xenotransplants to humans.
- ♦ Moral aspects of using animals for experimentation.

4. As a class, identify some preliminary goals and options for The Decision. What information might be needed to make an informed decision? Students should keep notes during this discussion. For example, one of the problems of medical research is the lack of certainty of outcomes and the high degree of risk involved in certain procedures. If time permits, students can gather information on the outcomes of xenotransplants, and the chances of disease transmission between animals and humans.

   An important concept is the genetic relationship between species and their vulnerability to each other's viruses. For example, humans are much more likely to contract a baboon virus than a pig virus because baboons are more closely related genetically to humans than pigs.

5. For homework, have students conduct research using the ideas generated in the class discussion to guide them. They should create a decision chart with the appropriate goals, options, and outcomes. The chart on page 39 is only an example. If students are familiar with probability, have them make probability estimates for the outcomes.

Part B

1. Conduct a general classroom discussion on the research results. What facts turned out to be most pertinent to The Decision? What are some value-based issues which science cannot resolve? Do not try to resolve the moral issues, but focus on how science could improve understanding.

2. Divide students into groups and have them use their decision charts to weigh the options using importance bars. Have students decide as a group.

3. In this activity, students act as government officials making a public policy decision about xenotransplants. Thus, their decision is based on a broad, social perspective. Ask students how the process

NOTE
Use Teaching Plan: Guided Activities (page 34) to structure the class work.

	Allow experimental heart xenotransplants to humans	Do not allow experimental heart xenotransplants to humans
Reduce human deaths	*Experimental transplants will not help the vast majority of transplant patients on the waiting list. However, advances in research could eventually lead to widespread use of animal transplants, and therefore save thousands of human lives in the long run.*	*This will not only fail to help people on the transplant wait list in the short run, but it will also fail to help larger numbers of people in the future because the procedure is not given the opportunity to succeed.*
Control costs	*Experimental surgeries and follow-up care are costly.*	*Money will be saved by not carrying out experimental surgery.*
Restrict use of animals for experimentation	*A greater number of animals will be used as experimental surgery increases.*	*Fewer animals will be used.*
Prevent virus transfer from animal to human population	*This poses a possible, but uncertain, threat to the person receiving the organ, and possibly to many other people.*	*Not allowing experimental surgery minimizes the potential for transmission of an animal virus to humans.*
Respect ethical/moral views	*The transfer of animal hearts to humans is not acceptable to certain religions. However, other people think that impeding medical research is immoral.*	*Some people will want the experimental transplants because they have loved ones who are sick and need a new organ.*

SAMPLE DECISION CHART: Xenotransplants

would change if they were making a similar decision from a personal perspective. For example: Your critically ill family member has been on the wait list for a heart transplant. A xenotransplant may save his or her life. Considering the risks and benefits of the procedure, do you agree to have it done?

Questions for discussion: Does changing the perspective from committee to family member influence your goals? Would it change the way you assigned advantages? How objective could you be in making a final decision? Would emotions bias your decisions? How does this compare with your feelings and personal opinions when first analyzing the issue? Compare how time would affect the decision for both scenarios.

4. For homework, have students write a brief (one or two paragraph) essay in their journals about what they learned.

Extension

A main reason for xenotransplants is the lack of human donors. Since the people who donate organs (except for some kidney donations) are generally not alive, the decision is up to family members. A few religions forbid the altering of the human body, and therefore forbid the extraction of organs from a dead body. Have students consider ways that more human organs could be obtained. An alternate decision could be: Should a public policy be instituted to obtain more voluntary donations from family members? Should hospitals pay the family of the donor for the organ?

Assessment

Review the students' journals using Appendix D. Use the Extension to evaluate student ability to apply learning, or have students choose and research their own bioethics issue.

Resources

Committee on Xenograft Transplantation, Institute of Medicine. 1996. *Xenotransplantation: Science, Ethics, and Public Policy.* Washington, DC: National Academy Press.

Science and *Science News* have been following developments in xenotransplants.

Internet sites

Animal Rights Resource Site
http://envirolink.org/arrs/essays/xeno_risks.html
An essay outlining some of the arguments against xenotransplants (not just animal rights).

University of Pittsburgh Medical Center: Xenotransplantation Program
http://www.upmc.edu/news/xenotxbg.htm
Descriptions of the history of xenotransplants.

U.S. Food and Drug Administration (FDA)
http://www.fda.gov/
Main FDA site.
http://www.fda.gov/opacom/backgrounders/xeno.html
FDA background paper on xenotransplants.

WhyFiles: Cross-Species Transplants
http://whyfiles.news.wisc.edu/007transplant/
Good student-level introduction to the issue.

Immutations

Background

During your lifetime you will have several immunizations. Most scientists and medical practitioners consider immunizations—often referred to as *shots*—to be among the greatest medical advances in human history. Before the development of immunizations, diseases such as smallpox, diphtheria, and polio were responsible for millions of deaths. The creation of vaccines for these diseases has practically eliminated their threat in many parts of the world. Despite the numerous benefits of vaccines, they are not without risks. Some people react to vaccines and become ill or die. Decisions concerning vaccines are not as simple as you might think.

Swine Flu: A true story

In January, 1976, soldiers at Fort Dix, New Jersey, came down with severe cases of what appeared to be the flu. One soldier died. Unsure of the exact nature of the disease, officials sent blood samples to the Centers for Disease Control and Prevention (CDC) in Atlanta, Georgia. In mid-February, the unknown virus was identified as the Swine Flu virus.

Swine Flu is a variant of the A/Victoria strain, which was responsible for the deaths of more than 500,000 Americans and over 20 million others worldwide in 1918 and 1919. Officials feared a similar outbreak would occur again. Additionally, if millions of people sick with the flu missed work, it could cost billions of dollars in lost productivity.

While officials closely watched the flu outbreak at Fort Dix, their concerns turned to the next winter's flu season. The director of CDC called a meeting in March, 1976 to determine the best course of action. The committee recommended two options: either immediately begin a large-scale vaccination effort; or store large quantities of Swine Flu vaccine while waiting to see whether an epidemic emerged.

An epidemic could be headed off by going forward with a large-scale vaccination program. However, such an effort would cost well over $100 million dollars. Furthermore, many vaccine-related illnesses and even deaths (from allergic reactions) could result from a widespread vaccination effort. Experts were not sure how many vaccine-related deaths might occur.

Storing the vaccine and waiting to see how the disease developed could save money and resources. Those favoring waiting claimed that the evidence of an epidemic was not yet strong enough to warrant committing large amounts of money. Yet some officials felt that at some point it might become too late to stop the spread of this disease.

The Decision

In mid-March of 1976, the committee presented these options to President Gerald Ford. Assume that you are President Ford and must make this difficult decision: Should the Swine Flu vaccine be widely distributed?

Additional Facts

March 24: President Ford announces the largest immunization campaign ever attempted in the United States. Starting in the fall, over 200 million people are to be vaccinated over a period of three months. Ford's decision is supported by some of the leading disease scientists of the day, including Jonas Salk, who developed the polio vaccine, and Albert Sabin, the creator of the oral polio vaccine.

October 1: The first vaccines are administered. Within ten days, more than 10 million people receive the vaccine.

October 11: Three people die shortly after receiving the vaccination. All are over 70 years old. During the next several days, 10 to 12 deaths occur daily for every 100,000 people between the ages of 70 and 74 years old. Despite concern, officials claim this is consistent with average death rates for this age group.

November 12: A Minnesota doctor finds a rare condition of the nervous system called Guillain-Barre (GB) syndrome. GB syndrome is an illness resulting from the inflammation and destruction of the myelin sheath covering nerve fibers. Symptoms include numbness in the fingers or toes and muscle weakness. Recovery usually occurs over a period of months with some patients requiring long-term rehabilitation. Permanent damage occurs in about 10 percent of cases, with mortality rates of three to four percent.

December 14: CDC receives reports of 54 GB cases, with one fatality. Of the 54 cases, 32 had been vaccinated for Swine Flu in the past month.

December 16: Over 40 million people have been vaccinated.

December 17: CDC abruptly suspends the vaccination program. New data show that people who receive the Swine Flu vaccination are seven times more likely than non-immunized people to develop GB syndrome.

Individual Risk of Swine Flu and Vaccine*

Vaccine

Probability of developing severe reaction:	2% to 10%
Fatal reaction (unknown but presumed):	< .0013%
Contracting GB syndrome:	.0013%

Swine Flu

Probability of contracting:	2.7%
Death from flu:	.0024%
Contracting GB syndrome:	.00025%

*The probabilities provided are for healthy individuals who are between the ages of 5 and 70.

Immunizations

Student Handouts

Student section (pages 41–42)
Appendix B: Summary of Decision Making
Appendix C: Decision Chart

Objectives

Students will:

♦ Learn about the Swine Flu debate of 1976.
♦ Create decision charts based on the information and come to a decision.
♦ View the results of the actual decision over time and reconsider the options involved.

Terms for Understanding

Immunization/vaccination, virus, epidemic, smallpox, diphtheria, polio

Procedure

Part A

1. Have students read "Immunizations" (student section). Do not hand out page 42 until the students have completed Part A. After a brief discussion of the topic, students should write a preliminary decision in their journals.

2. Discuss The Decision (in student section). Whose decision is it? This real-life scenario gives your students the opportunity to consider the decision President Gerald Ford made in 1976 during the Swine Flu outbreak. The case study points out difficulties in making decisions when there are unknown elements. Stress that the option to wait and see to gather evidence is not just avoiding the problem.

3. Risk assessment. Discuss the issue based on risk space (new vaccine, many unknown elements). Have students brainstorm some major issues in The Decision.

4. As a class, identify some preliminary goals and options for The Decision. What information might be needed to make an informed decision? Students should keep notes during this discussion.

5. For homework, have students create a decision chart with the appropriate goals, options, and outcomes. The chart below is only an example. To prepare for class, they should make a decision based on their chart.

6. Conduct a general classroom discussion on The Decision. What facts turned out to be most pertinent? What are some value-based issues which science cannot resolve? Do not try to resolve the moral issues, but focus on how science could improve understanding.

NOTES

♦ *Use Teaching Plan: Guided Activities (page 34) to structure the class work.*

♦ *First, students complete Part A. Then, hand out additional information for Parts B and C (page 42).*

SAMPLE DECISION CHART: National Swine Flu Immunization

	Begin immediate Swine Flu vaccination program	Wait and see
Eliminate work absenteeism and prevent lost productivity	*The risk to the economy is greatly reduced (less medical costs and work absenteeism).*	*The economic risks are greater.*
Control costs of implementing option	*High cost: $100 million.*	*Low cost initially, but cost of epidemic could be even greater than $100 million.*
Prevent death from flu	*Deaths would be reduced.*	*Potentially, there could be millions of deaths.*
Prevent vaccine-related illness or death	*There would be a relatively low number of deaths.*	*No risk.*

TEACHER SECTION

Part B

1. After students have reached a decision based on Part A information, hand out page 42. Conduct a class discussion on how each piece of information may change the class's original analysis.

   Questions: At what point should the vaccination program have been discontinued? Or should it have continued? Were careful estimates of the degree or magnitude of a potential Swine Flu epidemic considered? Would the 1976 epidemic necessarily have been as deadly as the 1918 epidemic?

2. Read the following to students:
   This scenario illustrates the problem of uncertain outcomes. Neither the president nor you could have predicted the risks or costs of the epidemic, or have knowledge of the actual risks associated with the vaccine. A decision was made in the absence of complete knowledge.

   Questions: Is there any way you or the president could have made a better decision? What are the limitations in conducting research to determine outcomes? Is it ever possible to know or predict all the outcomes when making a decision?

   Stress to students that good decision making doesn't mean that every outcome will be positive. A perfect decision analysis can still lead to negative outcomes.

Part C

1. Change the decision to the individual viewpoint. Tell your students they must choose between receiving a Swine Flu immunization or possibly contracting the flu virus. What are the risks involved in choosing or choosing not to get the vaccine?

2. Have students create a decision chart for their individual decisions. The chart below is only an example.

3. Have students assign values to the goals boxes in the decision chart. Contracting GB syndrome may be 20 times worse than being uncomfortable, but absenteeism may be only five times worse than being uncomfortable. For example, the relative values assigned to the goals could be: Pain from shot or discomfort from flu (- 1), Death (- 100), Absenteeism (- 5), GB Syndrome (- 20).

4. After assigning relative values to goals, students can consider the probabilities in their handout (Part C) for a single flu season. Discuss with students how younger or older people may have different probabilities, and how these may change the decision.

5. Students can compute expected values by multiplying the probability in each outcome box by the corresponding goal values. Lower values are less desirable. In the example chart on page 45, the decision is difficult to make because uncertainty gives a range of values for receiving the vaccine. If the relative values are assigned differently, then the final decision may change.

6. For homework, have students write a brief (one or two paragraph) essay in their journals about what they learned.

SAMPLE DECISION CHART:
Individual Swine Flu Immunization

	Receive vaccine	Do not receive vaccine
Avoid pain/discomfort	*You will have the discomfort of receiving a shot, and you might react to the vaccine.*	*The flu is unpleasant and painful.*
Avoid death	*Death is extremely unlikely.*	*Death from the flu is not a significant concern for a young, healthy person.*
Prevent work absenteeism	*Absenteeism is less likely.*	*You are more likely to miss school or work.*
Avoid GB syndrome	*You are slightly more likely to contract GB syndrome.*	*You have an extremely low chance of contracting GB.*

NATIONAL SCIENCE TEACHERS ASSOCIATION

	Receive vaccine	Do not receive vaccine
Pain/discomfort of flu or reaction	$-1 \times (.02$ to $.1) = -.02$ to $-.1$	$-1 \times .027 = -.027$
Death from flu	$-100 \times <.000013 = < -.0013$	$-100 \times .000024 = -.0024$
Absenteeism	$-5 \times (.02$ to $.1) = -.1$ to $-.5$	$-5 \times .027 = -.135$
GB syndrome	$-20 \times .000013 = -.00026$	$-20 \times .0000025 = -.00005$
Expected Value of each option =	**-.12 to -.60**	**-.16**

SAMPLE DECISION CHART:

Immunization vs. Flu Probabilities

Extension

Many parents are choosing not to immunize their children for diseases such as tuberculosis and measles. The risk of contracting these diseases is relatively small in the United States. However, measles cases did increase in the U.S. between 1989 and 1991 because some parents did not have their children immunized. Schools and universities sometimes require students to be completely immunized for enrollment. Should parents have their children immunized even if there is a slight risk of reaction to the vaccine?

Assessment

Review the students' journals using Appendix D. Assign the Extension to evaluate student ability to apply learning, or have students choose and research their own bioethics issue.

Resources

Copp, Newton. 1989. *Vaccines: An Introduction to Risk.* Stony Brook: Research Foundation of State University of New York.

Internet sites

Centers for Disease Control and Prevention (CDC)
http://www.cdc.gov/
Main CDC site.
CDC National Immunization Program
http://www.cdc.gov/nip/
General immunization information.

National Foundation for Infectious Diseases
http://www.medscape.com/Affiliates/NFID
Excellent fact sheets on a range of diseases requiring immunization.
http://www.medscape.com/Affiliates/NFID/library/general.html
See the infectious disease index.

U.S. Food and Drug Administration
http://www.fda.gov/
Main FDA site.
http://www.fda.gov//fdac/features/095_vacc.html
See paper: "How the FDA Works to Ensure Vaccine Safety."

Household Cleaning Products

Background

When we say that an object is "clean," we can mean two things. Clean can mean visibly free from dirt or residues. Clean also means disinfected, or free from disease-causing bacteria or viruses. Despite advertisements depicting killer bacteria running rampant, most surfaces in our homes do not need to be disinfected. Our skin is a protective layer which keeps out foreign organisms. If bacteria get past the skin into our bodies, our immune systems are able to fight the infection. Also, bacteria grow only in certain favorable conditions. Dry and visibly clean surfaces inhibit bacterial growth.

Two particular cases should receive special care: handling animal products, and cleaning up animal and human wastes. In these cases, some kinds of bacteria can cause illness if precautions are not taken. When handling uncooked food, especially meats and eggs, wash the surfaces the food contacted with warm soapy water, and then wash your hands. When cleaning up human or animal wastes, be careful not to contact areas where you might place food, and wash the surfaces and your hands in warm, soapy water. If necessary, a mild bleach solution will disinfect non-porous surfaces.

Some people with impaired immune systems should be especially careful. But, for generally healthy people, it is not possible, practical, or necessary to disinfect every surface of our homes.

Some commercially prepared cleaning materials are very powerful and convenient. You need only apply the substance and wipe with a cloth with no scrubbing, no rinsing required. Traditionally, cleaners were made of mixtures of water and vinegar, baking soda, or salt. All these are less convenient because one must rinse with water or sometimes scrub to remove dirt.

However, preventive maintenance can reduce some major cleaning tasks because dirt doesn't have the chance to build up. For example, if you wipe up spills immediately they usually don't require more than soap and water. If you clean more often, you may not need to scrub or use powerful cleaning solutions.

Advertisements suggest that we need a different cleaner for each task—from cleaning the toilet to the floor. Some materials do need special solvents, but most surfaces in the home need only an all-purpose cleaning solution. Some products are specialized and may be more efficient than an all-purpose cleaner at dissolving particular substances. You would have to experiment to see if this is true.

Some common cleaning solutions are hazardous—ignitable, explosive, radioactive, corrosive (capable of eating away materials or destroying living tissue), or toxic (poisonous, whether immediately or chronically). The United States Government requires labels on hazardous substances. The three levels of hazardous labeling on common household products are:

♦ Danger: substances which are extremely flammable, corrosive, or highly toxic;

♦ Poison: substances which are highly toxic;

♦ Warning or Caution: substances which are moderately or slightly toxic.

The labels only point out acute effects, such as poisoning, if you ingest the substance. Labels are not required to provide information on chronic effects, such as increasing the chances of cancer or birth defects. The word "nontoxic" on a label has no scientific definition according to U.S. laws. If you want to use only nontoxic products, either contact the manufacturer to get a list of ingredients, or make your own nontoxic cleaners.

Alternative products exist that clean most surfaces in the house but are relatively less toxic or even nontoxic. These substances are simple to use and almost always much less expensive. When buying a cleaning solution, look at the label. Some cleaning materials are combinations of less expensive products. For example, some commercial window cleaners include vinegar and water. You could make a less expensive cleaner with vinegar and tap water. Here are some nontoxic cleaning ideas:

Air fresheners

♦ Open the window.

♦ Sprinkle baking soda in odor-producing areas.

♦ Light a candle.

♦ Use soap and water on soiled, smelly surfaces.

Cleaning solutions for most surfaces

♦ Use water and one of the following: soap, baking soda, or vinegar.

Glass cleaners

♦ Wipe surface soil with a cloth, then use a vinegar and water mixture.

Oven cleaners

♦ Wipe up spill immediately.

♦ Pour salt or baking soda on greasy spill and wipe up later.

♦ Use a non-metallic scrub brush.

The Decision

You must clean your home each week. Pick a certain chore, such as cleaning the bathroom, kitchen, or living room, and decide whether or not to use a certain product. Keep in mind whether the product is hazardous, how much it costs, and if less expensive or less hazardous alternatives or techniques are available.

TEACHER SECTION

Household Cleaning Products

Student Handouts

Student section (pages 46–47)
Appendix B: Summary of Decision Making
Appendix C: Decision Chart

Objectives

Students will:
♦ Read about the differences between cleaning products in terms of cost, toxicity, and convenience.
♦ Decide what cleaning products to use in the home using decision charts.

Terms for Understanding

Disinfect, bacteria, immune system, solvents, hazardous, toxic, corrosive, ignitable, explosive, radioactive

NOTE
Use Teaching Plan: Guided Activities (page 34) to structure the class work.

Procedure

Part A

1. Have students read "Household Cleaning Products" (student section). Make sure students understand the difference between acute and chronic toxicity. Labels only warn against acute toxicity. After a brief discussion of the topic, students should write a preliminary decision in their journals.

2. Discuss The Decision. Whose decision is it? This decision is personal. Should it be made at a national level?

3. Risk assessment. Have students conduct a shopping survey. They should visit a store that carries cleaning products and compare labels and costs of various brands. They should also look for signal words such as: Caution, Warning, and Danger. Have the students use their data to develop a chart of the brands and their uses, whether they are labeled as hazardous, and how much they cost per use. Discuss whether they feel the cleaning products they surveyed are a risk in their home. Discuss the unit cost and active ingredient of the products. Are some more cost effective than others?

You could also have students conduct a cleaning experiment to compare the effectiveness of various materials. For example, compare the cleaning effectiveness of a commercially prepared glass cleaner to vinegar and water.

4. Have students identify some cleaners labeled as hazardous. Write this list on the board. Decide whether or not to use each one of these cleaners. If there are a number of questionable products mentioned, you may want to assign or have students choose one product. Students may work in groups or individually.

5. For homework, students should develop a decision chart for one product. Goals should include: cost, effectiveness, toxicity, and ease of use. See Resources for sources of information on the hazards of certain cleaning materials. For example, some oven cleaners contain sodium hydroxide, which is caustic and can burn skin or irritate mucous membranes if fumes are inhaled.

SAMPLE DECISION CHART: Household Products (Brand X Toilet Bowl Cleaner)

	Use commercial toilet bowl cleaner	Use alternative: scrub brush and baking soda
Effectiveness	*It cleans toilet bowls well.*	*This option is adequate, if scrubbing is done thoroughly.*
Low cost	*The product costs $ for X uses.*	*Baking soda costs $ for X uses.*
Ease of use	*It requires only easy scrubbing.*	*This is also easy, but necessitates more frequent cleaning than commercial cleaner.*
Low health risk	*This may be toxic.*	*This is nontoxic.*

Part B

1. Conduct a general classroom discussion on the research results. What facts turned out to be most pertinent to The Decision?

2. Divide students into groups and have them use their decision charts to weigh the options using importance bars. Have students decide as a group. Then, they should present their decisions to the class. For each product, ask students how their different goals affected the way they marked advantages and drew importance bars. What was the most important advantage?

3. For homework, have students write a brief (one or two paragraph) essay in their journals about what they learned.

Extension

Have students research wastewater treatment systems in your area. They can find out which products are filtered out of the water system. Some storm drain systems are not treated at all. Have students decide whether people should be able to wash their cars on the street if the water is not treated.

Assessment

Review the students' journals using Appendix D. Assign the Extension to evaluate student ability to apply learning, or have students choose and research their own environmental health issue.

Resources

Gosselin, Robert, et al. 1984. *Clinical Toxicology of Commercial Products.* Baltimore: Williams & Wilkins.

The Environmental Health Clearinghouse
(sponsored by the National Institutes of Health)
2605 Meridian Parkway, Suite 115
Durham, NC 27713
Phone: (800) NIEHS-94
Internet site: http://infoventures.com/e-hlth/
e-mail: envirohealth@niehs.nih.gov
Call or e-mail for free packet of background materials on household toxins and alternatives.

Internet sites

Centers for Disease Control and Prevention:
National Center for Infectious Diseases
http://www.cdc.gov/ncidod/ncid.htm
NCID: Foodborne Diseases
http://www.cdc.gov/ncidod/diseases/foodborn/
foodborn.htm
Background on causes and prevention of foodborne diseases.

Ozone

Background

What is ozone? Ozone is a chemical compound made up of oxygen. But most oxygen in the air exists as O_2, a relatively stable compound. Instead of two oxygen molecules, ozone has three oxygen molecules (O_3).

Where can you find ozone? Ozone has both positive and negative effects, depending on where it is found. When concentrated in the troposphere—the bottom layer of Earth's atmosphere where we live—it poses a health problem. This "bad" ozone is a major component of smog, produced by cars and other sources of combustion.

About 90 percent of ozone is found in the stratosphere, 10 to 50 kilometers above Earth's surface. Stratospheric ozone plays an important role in absorbing the sun's ultraviolet (UV) radiation. Life has evolved on Earth with a certain level of ozone and a certain amount of UV radiation. Natural variations in this level have occurred in such a way that evolutionary changes have been able to compensate. The changes caused by industrial technology are much more rapid than evolution in some species can handle. Too much UV radiation can increase the incidence of skin cancer, cause crop damage, and drastically affect phytoplankton—the base of the ocean's food chain. The proper balance of ozone in the stratosphere keeps these problems from occurring.

Effects of ozone

In 1929, researchers developed chlorofluorocarbons (CFCs) for use as a refrigerant. Before CFCs, ammonia and other poisonous or highly flammable chemicals had been used for this purpose. CFCs make excellent coolants for refrigerators and air conditioners because they are nontoxic, noncorrosive, nonflammable, and unreactive with most substances. They also work well in foam cups and plates to trap heat. They were used as refrigerants, insulation, and in aerosol sprays. By the mid-1970s, the use and production of CFCs was a global, two billion dollar per year industry.

In the early 1970s, a high speed aircraft called the supersonic transport, or SST, was built to transport people faster than the speed of sound. Concerns about emissions at high altitude led researchers to explore the effects of high amounts of nitrogen and other chemicals on the stratosphere. The U.S. development of the SST was canceled in part due to environmental concerns. Prompted by these concerns, scientists began to study other possible human-caused influences on atmospheric ozone.

In the 1970s, chemists F. Sherwood Rowland and Mario Molina began to calculate the reactions of ozone to CFCs. They knew that CFCs had the potential to destroy ozone compounds, and extended their thinking to what might be happening to all the CFCs that were entering the atmosphere. They reasoned that CFCs were destroyed only when they rose into the stratosphere and were broken down by the sun's UV light.

In the stratosphere, sunlight breaks down CFCs into chlorine. While CFCs don't directly destroy ozone—they are actually chemically inert—the chlorine that comes from their decomposition by the sun's UV light does destroy ozone. For each atom of chlorine present in the stratosphere, two reactions occur:

$$Cl + O_3 >>>> ClO + O_2 \quad and \quad ClO + O >>>> Cl + O_2$$

The overall reaction results in the breakdown of ozone into oxygen. Chlorine acts as a catalyst, drastically increasing the rate of conversion of ozone to oxygen. The chlorine atom is a reactant in the first reaction, but is regenerated as a product in the second reaction, thus acting as a catalyst. Because chlorine is not consumed in the réaction, it is possible for one chlorine atom to destroy 100,000 ozone molecules.

You may wonder why ozone—considered a pollutant in the troposphere, or lower atmosphere—doesn't replenish stratospheric ozone. Of Earth's ozone, 90 percent is in the stratosphere, so the tropospheric air entering the stratosphere has less ozone than the stratospheric air already there. In fact, the stratosphere is a source of ozone for the upper troposphere.

You may also wonder why CFCs don't help combat smog in the troposphere, since their chlorine catalyst contribution works so well in the stratosphere. They only contribute their chlorine catalyst when broken down by the Sun's high energy UV radiation in the upper atmosphere. They are chemically inert near Earth's surface.

Alternatives to CFCs?

From the 1970s to the present, industries have researched alternatives to CFCs. But each of the new options, like CFCs, has trade-offs. Some may be more expensive, some a little less efficient, or some may be toxic.

The Decision

The year is 1985. The British Antarctic Survey, which had been monitoring atmospheric properties over Antarctica since 1956, found stratospheric ozone to have significantly decreased—to half of its normal distribution above Antarctica. Despite this alarm, scientists had no substantial proof pointing to the direct cause of the depletion. They proposed three hypotheses to explain the depletion of ozone over Antarctica: (A) fluctuations in solar intensity cause changes in the natural atmospheric chemistry, resulting in decreased levels of ozone; (B) changes in wind patterns cause decreases in ozone; (C) increasing amounts of CFCs provide more chlorine to serve as a chemical catalyst to break down ozone.

You are an advisor to the President of the United States in the 1980s. At this point, it is clear that CFCs contribute to the ozone depletion problem, but scientists are not completely sure that it is the only causal factor. Will you recommend a ban on CFCs? Do you have other options?

Questions

Read the following information and answer the questions.

By 1987 research uncovered possible answers to the questions posed by the Antarctic researchers:

(A) Solar intensity was determined not to alter atmospheric chemistry or result in decreased levels of ozone.

(B) Wind patterns were found not to decrease ozone in the stratosphere.

(C) Chlorofluorocarbons were linked to the ozone depletion over Antarctica. This link was found by sampling chlorine at different latitudes and altitudes in the stratosphere. In the stratosphere over Antarctica, abnormally high levels of chlorine were found. This chlorine was not found in high concentrations in other parts of the atmosphere. Researchers refer to this discovery as the "smoking gun," because it proved that chlorine was the culprit in ozone depletion. The majority of evidence now shows that the source of chlorine was chlorofluoro-carbons.

Questions: *Would the new information affect your decision? Would you now institute an immediate ban on CFCs? Explain your answer.*

On September 16, 1987, the United Nations adopted the Montreal Protocol on Substances that Deplete the Ozone Layer. The Protocol called for a drastic reduction in chlorofluorocarbons. Many industrialized countries, including the United States, completely banned the harmful compounds. Some developing countries did not support the ban.

Despite efforts to curb CFC emissions, stratospheric ozone continues to deteriorate throughout the world. In 1987, the ozone hole over Antarctica was two times the size of the United States. In 1992 and 1993, the lowest levels of ozone were recorded.

Question: *Why do these levels continue to decline despite efforts to curb CFC emissions?*

The World Health Organization estimates that 1.6 million cases of glaucoma and 100,000 new cases of skin cancer will develop each year as a result of increased UV exposure. This is in part due to ozone depletion.

Questions: *What type of precautions can be made to avoid these medical problems? What should you do to avoid these problems? What personal actions can you take to prevent further ozone deterioration?*

Ozone ◆ TEACHER SECTION

Student Handouts

Student section (pages 50–52)
Appendix B: Summary of Decision Making
Appendix C: Decision Chart

Objectives

Students will:

◆ Learn the difference between ozone in the troposphere and ozone in the stratosphere.
◆ Develop an understanding of the causes of ozone depletion in the stratosphere.
◆ Make a decision whether or not to ban CFCs based on information available to scientists and policy makers at different points in the ozone research process.

Terms for Understanding

Ozone, chlorofluorocarbons, troposphere, stratosphere, ultraviolet radiation, smog, phytoplankton, inert, catalyst

Procedure

Part A

1. Have students read "Ozone" (student section). Note: Do not hand out page 52 until after students have completed Parts A and B. Discuss the chemical nature of ozone, different parts of the atmosphere, and the difference between surface and stratospheric ozone. Then, students should write a preliminary decision in their journals.

2. Discuss The Decision. Whose decision is it? Discuss the economic implications of banning CFCs. Assuming it is 1985, ask students what this could mean for the two billion dollar per year chlorofluorocarbon industry. Students should consider this from the point of view of both producers and consumers. Remember that there are benefits to CFCs.

3. Risk assessment. Rowland and Molina speculated that chlorine derived from the break-up of CFCs may be responsible for ozone depletion. Ask students to theorize on other possible sources of chlorine. Does chlorine from these sources reach the stratosphere?

4. As a class, identify some preliminary goals and options for The Decision. What information might be needed to make an informed decision? Students should keep notes during this discussion.

5. For homework, have students conduct research using the ideas generated in the class discussion to guide them. They should create a decision chart with the appropriate goals, options, and outcomes. The chart on page 54 is only an example. If students are familiar with probability, have them make probability estimates for the outcomes.

Part B

1. Conduct a general classroom discussion on the research results. What facts turned out to be most pertinent to The Decision? What are some value-based issues which science cannot resolve?

2. Break students up into groups and have them use their decision charts to weigh the options using importance bars. Have students decide as a group.

3. Lead a general class discussion of the group decisions. Try to identify what areas could be resolved through scientific research and what areas are moral/values issues.

Part C

1. This scenario demonstrates the nature of decision making for a long-range problem. Discuss the role of continuing research and the need to reevaluate decisions. For example, have students research current ozone levels and scientific predictions. Can they think of other long-range problems that require ongoing evaluation? Some examples are global warming, topsoil erosion, and deforestation.

2. For homework, assign the questions in the student section (page 52).

Extension

Consider other issues relating to environmental chemistry in which decisions have to be made about potentially harmful chemicals. Examples might include sulfur emissions, PCBs dumped into aquatic systems, and automobile emissions.

> **NOTE**
> Use Teaching Plan: Guided Activities (page 34) to structure the class work.

	Ban CFCs immediately	Continue use
Keep convenience and uses for CFCs	*Some replacements are available that combine convenience and safety of CFCs, but these also have risks and benefits.*	*Consumers would have the convenience of air conditioning, styrofoam, and refrigeration.*
Protect economy	*A ban would be very costly (loss of income to industries, cost of producing alternatives).*	*Continued use saves industry and jobs with little economic damage in short run.*
Prevent health problems	*A ban could prevent some health effects.*	*Health effects may be significant with higher rates of skin cancer.*
Protect plants and animals	*A ban could prevent damage from UV to plants and animals.*	*Plants and animals could be significantly harmed. Crop production may decrease, forests might be damaged, and certain animal populations might be reduced.*
Control cost of continued research	*A ban on CFCs would perhaps decrease research money spent on ozone depletion, although monitoring efforts will continue.*	*Rigorous and expensive research will need to continue.*

SAMPLE DECISION
CHART: CFCs

Assessment

Review the homework assignment using Appendix D. Assign the Extension to evaluate student ability to apply learning, or have students choose and research their own environmental chemistry issue.

Resources

Albritton, Daniel, et al. 1992. *Our Ozone Shield.* Reports to the Nation On Our Changing Planet. no. 2 (Fall). Boulder, CO: University Corporation for Atmospheric Research.

Internet sites

U.S. Environmental Protection Agency
http://www.epa.gov/
Main EPA site.

EPA: Stratospheric Ozone Program
http://www.epa.gov/ozone/index.html
Best overall source on the topic. Includes many links monitoring and policy organizations and to industries developing substitutes for CFCs.

National Oceanographic and Atmospheric Administration
http://www.noaa.gov/
Main NOAA site.
http://www.ogp.noaa.gov/OGPFront/mono2.html
On-line version of Our Ozone Shield.

Groundwater

Background

Water is an important part of our lives. We *are* water; it makes up much of our bodies. We use it to wash and to make a wide variety of products. While Earth holds huge reserves of water, only a small amount is easily obtainable by humans. About 97 percent of Earth's water is salt water— available for humans only after great expense of energy. Of the three percent that is fresh water, 2.4 percent is frozen icecaps and glaciers, 0.4 percent is groundwater, .02 percent is surface water, and the remaining amounts are in air and soil. With such a small percentage of Earth's water readily usable, it is important to be concerned about its quality and protection.

Surface water is found in streams, rivers, lakes, reservoirs, and wetlands. Surface water supplies a large percentage of the world population with fresh water. Groundwater consists of water that sinks into the Earth where it can be stored for long periods of time. Communities are increasingly tapping groundwater as surface water becomes contaminated and more scarce. Half of all Americans depend on groundwater for drinking water.

Ground and groundwater are closely linked. As humans alter Earth's surface, the water in the ground may be contaminated. Groundwater contamination occurs when the chemical make-up of the groundwater is altered, either directly, due to the spill or leakage of a liquid, or indirectly, through the alteration of the ground through which water passes. High concentrations of chemicals can reach levels that render water unsuitable for human use.

Sources of groundwater contamination can be classed as either point sources or non-point sources. For example, point source pollution can come from:
- Improperly designed or maintained landfills for solid non-hazardous municipal waste or for hazardous wastes.
- Leaking underground storage tanks. There are about a million underground storage tanks in the U.S.; since the mid 1980s over 315,000 have been confirmed to leak.
- Leaking sewer and petroleum pipelines.
- Improper discharge from sewage and household wastes from septic tanks. There are 20 million septic tanks in the U.S.
- Leaks from mining wastes, such as in tailings ponds.
- Leaks from sewage lagoons and rapid infiltration ponds for liquid municipal waste. In all of the above examples, waste fluids, or leaches formed from the passage of rainwater and snowmelt through the waste, can flow across the unsaturated zone of the Earth's crust and deliver contaminated water to the water table.

Non-point sources of groundwater contamination can include:
- Agricultural and residential uses of herbicides, pesticides and fertilizers.
- The use of salts and other de-icers to clear the roads.
- Spills or slow drips of oils and anti-freezes from vehicles. Increased concentrations of vehicles may accelerate this process.
- Runoff of animal wastes from livestock areas.

Understanding the link between the human activities on the surface and the water below the surface is fundamental in preserving groundwater supplies.

The Decision

Part A

Analyze your community's groundwater. You need to determine how vulnerable your local groundwater is to contamination. The factors you should evaluate are:

♦ Depth: the distance from the ground surface to the water table.
♦ Precipitation: the rate of precipitation per year.
♦ Geology: the type of rock in which the aquifer is located.
♦ Soil: the type of soil lying above the aquifer.
♦ Topography: the slope and land surface located above the aquifer.

TABLE A:
Survey Your Aquifer

	Description	Value
Depth		
Precipitation		
Geology		
Soil Type		
Topography		

TABLE B: Vulnerability Factors for Aquifers

	Value	Definition
Depth	3	*If less than 3 meters of ground lies between surface and water table*
	2	*If between 3 meters and 25 meters*
	1	*If greater than 25 meters*
Precipitation	3	*If more than 1 meter of rain falls each year*
	2	*If between 38 and 100 centimeters*
	1	*If less than 38 centimeters*
Geology	3	*If rock is consolidated sediments*
	2	*If rock is sedimentary (sandstone, limestone)*
	1	*If rock is igneous or metamorphic*
Soil Type	3	*If soil is sand or gravely*
	2	*If soil is loamy—a mix of sand, clay and plant and animal matter—or silty*
	1	*If soil is clay-like*
Topography	3	*If land surface is mostly flat*
	2	*If land surface consists of gently rolling hills*
	1	*If land surface consists of steep hills or mountains*

VULNERABILITY SCALE

Sum of Factors	Vulnerability
5.0	Extremely Low
>7.5	Low
>10	Average
>12.5	High
15	Extremely High

Part B

Assume that your community is in great need of good stores. People in the community travel to other communities to do their shopping. A proposal has come forth to build a shopping mall with many stores. The mall would provide jobs for young people and adults. It would also encourage the creation of more restaurants and gas stations surrounding the mall. Assume the role of a city planning board and develop a decision chart to decide whether or not a mall should be built.

Groundwater

Student Handouts

Student section (pages 55–57)
Appendix B: Summary of Decision Making
Appendix C: Decision Chart

Objectives

Students will:

♦ Learn different factors that influence groundwater vulnerability.

♦ Gather data on their community's groundwater vulnerability.

♦ Develop decision charts based on a hypothetical scenario that has the potential to threaten groundwater in their community.

Terms for Understanding

Aquifer, groundwater, contamination, herbicides/pesticides, loam, topography, sedimentary, igneous, and metamorphic rock

Procedure

Part A

1. Have students read "Groundwater" (student section). Discuss whether and how groundwater is used in your community. Then, students should write a preliminary decision in their journals.

2. Discuss The Decision (Parts A and B). Part A is a survey of your local aquifer. Part B is a hypothetical decision on building a mall.

3. Risk assessment: To approach your local groundwater evaluation, you'll need to discover something about the geology and meteorology of your community. Your class's field studies in Earth science (even on the school grounds), agricultural extension agents, local meteorologists, engineering geology firms, and local university geology departments are all potential resources.

♦ How deep is your water table, or how deep do your community's wells need to be? (This will vary, but you should be able to find typical numbers. Why do some wells need to be deeper than others?)

♦ How much precipitation does your community receive in a year?

♦ What is the geology of your community? What rock type lays under the soil?

♦ What kind of soil does your community have?

♦ What is the topography of your area?

4. Have students record the answers to the above questions in Table A in the student handout. After students have gathered the data and placed it in Table A, they will use the values in Table B to evaluate the degree of vulnerability for your community's aquifer. The sum of the values should be compared to the vulnerability scale in the handout.

Questions: What is the meaning of the vulnerability descriptors? How vulnerable is the ground water in your community?

Part B

1. As a class, identify some preliminary goals and options for The Decision (Part B). Factors impacting the decision may include traffic, jobs, shopping, etc. Avoiding groundwater contamination must be one of the goals.

2. For homework, have students use the ideas generated in the class discussion to create a decision chart with the appropriate goals, options, and outcomes. The chart on page 59 is only an example.

Part C

1. Conduct a general classroom discussion on the research results. What facts turned out to be most pertinent to The Decision? What are some value-based issues which science cannot resolve?

2. Break students up into groups and have them use their decision charts to weigh the options using importance bars. Have students decide as a group.

3. Ask students to compare their final decision against their personal views and preferences. How do the city planning board's goals compare with student goals? Would the decision change if they were to consider the issue from a personal perspective?

NOTE

Use Teaching Plan: Guided Activities (page 34) to structure the class work.

	Build mall	Do not build mall
Control traffic	*Building will increase concentration of traffic around mall. Roads may be insufficient to handle traffic.*	*Traffic patterns will remain similar.*
Create jobs	*Mall will provide many jobs. Most of the jobs will be entry-level and custodial.*	*Job rate will not be altered.*
Improve shopping	*Additional stores will improve shopping.*	*Shopping opportunities will not improve, and people will need to drive other places to shop.*
Protect groundwater	*Groundwater may be threatened depending on data previously gathered, and location of mall.*	*No threats to groundwater will occur.*

Discuss the issue of tradeoffs with the class. Ask students what would happen if shopping and jobs carried greater weight than protecting groundwater. How then would you go about making the decision?

4. For homework, have students write a brief (one or two paragraph) essay in their journals about what they learned.

Extension

Students may consider another scenario which impacts groundwater, such as the effects of logging, increased agriculture, or more housing developments. Many areas of the country are facing these exact issues, and must consider tradeoffs which may damage natural resources.

Assessment

Review the students' journals using Appendix D. Assign the Extension to evaluate student ability to apply learning, or have students choose and research their own environmental issue.

Resources

Crowder, Jane and Cain, Joe. 1997. *Water Matters.* Vol. 2. Arlington, VA: National Science Teachers Association.

Kauffman, Sue Cox. 1994. *Water Matters.* Vol. 1. Arlington, VA: National Science Teachers Association.

The Groundwater Foundation
P.O. Box 22558
Lincoln, NE 68542
Phone: (402) 434-2740

Water Environment Federation
601 Wythe Street
Alexandria, VA 22314
Phone: (703) 684-2492
Internet site: http://www.wef.org/

Internet sites

U.S. Environmental Protection Agency: Office of Water Resources
http://www.epa.gov/OW/water.usgs.gov/

U.S. Geological Survey: Water Resources Division
http://water.usgs.gov/

The Politics of Biodiversity

Background

Think of all the different kinds of plants and animals that you have seen in nature, at the zoo, or on television. Estimates of Earth's total number of species range from 10 million to 100 million. We use the term *biodiversity* to refer to the number or diversity of species, as well as the habitats and ecosystems they comprise. For many reasons, greater biodiversity is better than less diversity.

Greater biodiversity allows nature to adapt more quickly to disturbances such as climate change. Earth's climate itself depends in part on the existence of diverse plant life, which moderates the atmospheric content. Humans depend on biological resources for food, clothing, shelter, and even medicines. More variety ensures that new products will be available and that one disease or calamity cannot wipe out all species at once.

Many people also feel that biodiversity has inherent value. Humans benefit psychologically from their awareness of the variety and uniqueness of the kinds of living things on Earth. Also, ecotourism is very popular these days because people like to see living plants and animals in their natural settings.

Today, many scientists believe that biodiversity is decreasing at the fastest rate in 65 million years. Estimates vary, but suggest that at least 20 percent of the world's species will be extinct within the next 25 years. The decline in biodiversity is largely the result of human activity and represents a serious threat to human development. Since the year 1600, the known causes of animal extinctions can be grouped as follows: habitat destruction (36 percent), over-harvesting and hunting (23 percent), the inappropriate introduction of foreign plants and animals (39 percent), and other causes (two percent).

When referring to the world's nations, we generally say a nation is more developed (such as the U.S.) or less developed (such as Rwanda). Most tropical ecosystems exist in less developed nations. Since most species (74 percent) are in these tropical ecosystems, many environmentalists think habitat preserves should be created in those countries.

Every nation needs to maintain independence from other nations, develop its own economy, and balance the use and conservation of its natural resources. Therefore, each nation must balance the ideal of biodiversity conservation with the immediate material and economic needs of its citizens.

The Decision

You are the president of a small, less developed nation. Fifty percent of your country is tropical forest habitat. Your people are generally poor farmers and live in rural areas on the edges of tropical forests. You must decide how best to allocate your national budget. Your advisors have given you the report below. The nation's currency is the "isla."

What is the best strategy to maintain biodiversity in your nation and still develop your economy?

State of the economy

♦ National debt: 1 billion islas
♦ Annual income from agriculture: 15 million islas
♦ Possible annual income from expanding agriculture by cutting more tropical forest lands: up to 30 million islas, but it would fall off after a few years as the soil lost its nutrients
♦ Current annual income from tourism: 2 million islas
♦ Possible future annual income from ecotourism: 5 million islas
♦ Possible future annual income from sustainable use of tropical forests: 5 million islas

Public opinion

Most of your people understand that conserving biodiversity is a good thing, but they also have pressing needs for jobs and resources to live. They might be convinced that conservation is important if it brought more jobs and a higher standard of living. Your citizens think you are a good president, but they are concerned that you might sell land to other nations to raise money. They feel that the nation's land should not be available to foreigners.

Debt-for-land offer

Your advisors tell you that an American conservation organization has worked out a deal with the World Bank. The group would pay 20 million islas for 1/4 of the nation's tropical forests in exchange for having 1/4 of your national debt to the World Bank forgiven. But this would mean that your land would become a nature preserve and never be available for development in the future.

The Politics of Biodiversity

Student Handouts

Student section (pages 60–61)
Appendix B: Summary of Decision Making
Appendix C: Decision Chart

Objectives

Students will:

♦ Learn some of the reasons for decreases in Earth's biodiversity.

♦ Assume the role of president of a small nation who must decide how to allocate resources for conservation and/or economic development.

Terms for Understanding

Extinction, biodiversity, habitat (habitat preserves), ecosystem, ecotourism

Procedure

Part A

1. Have students read "The Politics of Biodiversity" (student section). After a brief discussion of the topic, students should write a preliminary decision in their journals.

2. Discuss The Decision. Whose decision is it? Have students brainstorm some major issues in The Decision. Some examples are:

♦ The problems faced by political leaders in keeping public support.

♦ Methods to conserve biodiversity.

NOTE

Use Teaching Plan: Guided Activities (page 34) to structure the class work.

♦ Whether or not a small nation has the obligation to conserve biodiversity at the expense of its economic growth.

♦ Whether or not selling land to a foreign organization reduces a nation's sovereignty.

3. Risk assessment. Using their decision journals, have students do a risk assessment for loss of biodiversity and a risk assessment for maintaining political power in a nation.

4. As a class, identify some preliminary goals and options for The Decision. What information might be needed to make an informed decision? Students should keep notes during this discussion.

5. For homework, have students conduct research using the ideas generated in the class discussion to guide them. They should create a decision chart with the appropriate goals, options, and outcomes. The chart below is only an example. If students are familiar with probability, have them make probability estimates for the outcomes.

Part B

1. Conduct a general classroom discussion on the research results. What facts turned out to be most pertinent to The Decision? What are some value-based issues which science cannot resolve?

2. Break students into groups and have them use their decision charts to weigh the options using importance bars or expected value. Have students decide as a group.

SAMPLE DECISION CHART: The Politics of Biodiversity

	Focus effort on developing economy	Focus effort on biodiversity conservation
Maintain independence from other nations	*A strong economy would help maintain your independence.*	*Selling lands would give the conservation group a say in managing your country's resources now and in the future.*
Increase your nation's wealth	*Cutting forest lands would bring in agricultural revenues. Developing tourism and sustainable tropical products are alternatives for economic growth.*	*Selling forest to the conservation group would reduce the nation's debt and increase wealth.*
Maintain your political power	*A stronger economy would be popular because your people are poor.*	*If conservation reduces jobs and economic growth, you might lose power.*
Conserve biodiversity	*If you cut forest lands for agriculture, biodiversity will be reduced. Sustainable use of tropical forest would conserve biodiversity.*	*Conservation would preserve biodiversity.*

3. Lead a general discussion of the group decisions. How does the final decision compare to the preliminary decision? Are they different? Why or why not? What uncertainties make the decision difficult?

4. The United Nations has an environmental program that looks at ways to conserve biodiversity around the world. Have students reevaluate the biodiversity issue from the perspective of the United Nations. Have students consider if one country should pressure another to conserve its resources and how more developed nations could help less developed nations conserve habitat through funding programs. See Resources, below, for United Nations Internet site.

Questions for discussion: Should the United Nations be involved in international environmental decisions? What would be the best plan involving all nations to conserve biodiversity? How do your feelings change about biodiversity when its the Earth instead of a fictional country?

5. For homework, have students write a brief (one or two paragraph) essay in their journals about what they learned.

Extension

Quantifying biodiversity is a huge scientific challenge. We can only estimate the total number of species on Earth. Of the estimated 10 million to 100 million species of plants and animals, only 1,413,000 have actually been classified. Tropical forests may potentially hold millions of undiscovered species. In addition, species are difficult to track consistently, and counting just one species can take decades. The costs are also very high to conduct this kind of research. Since much of the biodiversity data is based on estimates, information is lacking. The issue of biodiversity points out some limitations of scientific studies and how decisions must be made in the absence of perfect information (see page 12). Have students decide: Should a nation have perfect biodiversity information before committing itself to conservation? How important is it to know how many species there are? How will a nation know if the conservation efforts are succeeding?

Assessment

Review the students' journals using Appendix D. Assign the Extension to evaluate student ability to apply learning, or have students choose and research their own environmental issue.

Resources

National Science Teachers Association. *Biodiversity*. 1997. Global Environmental Change Series. Arlington, VA: National Science Teachers Association.

National Science Teachers Association. *Deforestation*. 1997. Global Environmental Change Series. Arlington, VA: National Science Teachers Association.

Wilson, E.O. 1992. *The Diversity of Life*. New York: W.W. Norton & Company.

The Nature Conservancy
1815 North Lynn Street
Arlington, VA 22209
Phone: (800) 628-6860
Internet site: http://www.tnc.org/

Internet sites

EnviroLink
http://www.envirolink.org/
Environmental information resource site.

United Nations Environment Programme
http://www.unep.org/
Students can research the programs administered by the UNEP.

Speed Limits

Background

A fundamental law of physics is the relationship between momentum, mass, and velocity. Momentum is the product of an object's mass and its velocity. An increase in either mass or velocity, or both, corresponds to an increase in momentum. This means that a vehicle traveling at a higher speed will have greater momentum, and a crash will have greater severity.

Car crashes are more likely to be serious or fatal, and involve more property damage, as the speed of the car increases. Speed also reduces the driver's reaction time to avoid an obstacle, increases the distance needed to stop the vehicle, and reduces the effectiveness of safety features such as air bags, seat belts, and crumple zones in the car's frame.

Just because a car is going faster doesn't mean that an accident will happen. Car accidents depend on a great many factors including driver ability and impairment (such as intoxication), road conditions (wet or dry), type of road (highway versus two lane road), and population density (urban versus rural). For example, according to the National Highway Traffic Safety Administration (NHTSA), 77 percent of fatal accidents occur on two lane roads. And 88 percent of speeding related fatalities occurred on roads that were not interstate highways.

Effectiveness of speed limits

The effectiveness of speed limits to control traffic speed is somewhat debatable because several studies contradict each other. A federal study completed in June 1996, "Effects of Raising and Lowering Speed Limits on Selected Roadway Sections," indicated that most vehicles travel at the same speed along a given stretch of road—and that speed has little to do with the posted speed limit. The speed at which most cars travel is related to driver perception of a prudent speed to travel. When speed limits are raised, drivers do not significantly speed up. There was no evidence that lowering the speed limit reduced the accident rate or that raising the speed limit increased the occurrence of accidents.

Other studies based on recent increases in state speed limits on interstates have been mixed. In some state studies, when speed limits were raised the average speed did increase. According to these studies, people seem to pick a speed slightly higher than the speed limit no matter how fast the limit. But in other studies, especially one in Montana, the average speed on interstates did not significantly increase even when the speed limit was changed to any "reasonable and prudent" speed.

Speed limits provide valuable information to drivers about the existence of cross traffic, residential areas with children playing, school zones, and other situations that are not obvious.

Safety and speed limits

Reducing injuries and deaths in crashes is an important goal. Although actual numbers of fatalities change from year to year, comparisons of fatalities per vehicle miles traveled are the lowest in the history of the automobile. Although our society travels more miles per year by motor vehicle, we are less likely to die per mile traveled. Some reasons for the low fatality rate per mile traveled are improvements in car and road design, improved vehicle safety features, and more control over drunk driving.

Speed is an important factor in injury prevention because the higher the speed the more severe the crash. NHTSA cites "speed related factors" as contributing to one-third of all crashes. But other factors also contribute to crashes. For example, the four largest contributing causes to crashes are drinking alcohol, driving too fast for road conditions, failing to yield, and careless driving. However, in the case of driving too fast for road conditions, this may not mean just speeding over the posted speed limit, but speeding over what the conditions at that moment would require. Even if the speed limit is 55 mph on a certain road, on wet days a prudent driver would reduce speed. Therefore, posted speed limits are not the only factors in speed-related accidents.

Costs of accidents

The costs of crashes can be categorized as: property damage, productivity loss, medical, insurance administration, legal, funeral, and emergency service costs. The most costly crashes to society are those in which someone is injured. But the most common crashes are property damage only (86 percent of all crashes).

It makes sense to look at the causes of the most costly and severe accidents. Alcohol-involved crashes tend to be more severe and cause up to 30 percent of all crash costs. Exceeding the speed limit or driving too fast for conditions were also big factors in costs. However, as noted above, driving too fast for conditions is not necessarily related to the speed limit.

Using safety devices reduces the cost of accidents because they reduce injuries. Used properly, seat belts and airbags do reduce the overall injury and death rate in accidents. Again, these are less effective at high speeds.

Although accidents can be tragic and cost money to society, they contribute income to other parts of society: car repair shops, new car dealers, and medical practitioners. Although insurance companies do not make money on a car crash itself, the companies do make money from the premiums paid by the majority of people without auto accident claims.

Pollution and speed

Lower speeds reduce emissions—up to a point. A car that drives smoothly and slowly will pollute slightly less than a car that drives smoothly and quickly. However, a car that stops and starts in traffic will also emit a good deal of pollution.

If one assumes that cars are already traveling faster than the speed limit, raising the limit may not necessarily raise emissions. But if the average speed does increase with the raised speed limit, then emissions will increase.

The Environmental Protection Agency (EPA) has found that 10 percent of the cars on the roads emit 50 percent of the pollution from cars. These "gross emitters" are cars that are not tuned correctly or do not have catalytic converters. The majority of cars built since 1990 emit small amounts of pollutants at any speed.

Other strategies to reduce emissions could be more effective. For example, in some cities pollution has been reduced when the traffic lights were retimed to reduce stop and start traffic. Improving bus systems or adding bike paths reduces emissions by eliminating vehicles.

Revenues from speeding tickets

Many towns rely on revenues from motor vehicle tickets to pay for essential government functions without raising taxes. When speed limits are slightly lower than prevailing traffic, the police have more opportunities to issue tickets. Raising speed limits might reduce the number of tickets collected.

Other ways to collect ticket revenues could be to ticket people who do not use seat belts or to set up barricades to screen drivers impaired by alcohol. Random screenings, however, may be unconstitutional.

The Decision

You are a member of the town council of your suburban town. A certain busy stretch of road in town carries about 6,000 cars per day. The road runs through a business area of the town. Some citizens would like the speed limit raised because they say it's too low for road conditions. Your popularity might increase if you raised the speed limit. But another group of citizens wants even more traffic controls to control drunk and reckless driving.

You must decide whether to change the speed limit on this section of road. Your traffic administrator has given you some facts about the road. The current speed limit is 30 mph. The actual speed traveled by most cars on this road is 45 mph. About 12 speeding tickets are issued per day on that stretch of road. The average cost of the ticket is $75.00. The accidents on this road tend to occur when people are slowing and turning into businesses along the road. Drunk drivers also cause accidents, especially at night.

Speed Limits

Student Handouts

Student section (pages 64–66)
Appendix B: Summary of Decision Making
Appendix C: Decision Chart

Objectives

Students will:

♦ Apply the principles of momentum to vehicular accidents.

♦ Create a decision chart in an effort to decide if highway speed limits should be raised.

Terms for Understanding

Momentum, inertia, mass, velocity, crumple zones, emissions, catalytic converters

Procedure

Part A

1. Have students read "Speed Limits" (student section). Then, students should write a preliminary decision in their journals.

2. Discuss The Decision. Whose decision is it? The students are to assume the role of a town council member who must decide whether to raise the speed limit on a certain road. Major discussion points should include:

♦ The effect of speed limits on actual traffic flow.

♦ The costs to society of motor vehicle accidents.

♦ The contributing factors in accidents.

♦ The proper role of the government in reducing accidents.

3. Risk assessment. Discuss how the public sees the issue of speed limits based on risk space (familiar risk, large degree of control).

4. As a class, identify some preliminary goals and options for The Decision. What information might be needed to make an informed decision? Students should keep notes during this discussion.

5. For homework, have students conduct research using the ideas generated in the class discussion to guide them. They should create a decision chart with the appropriate goals, options, and outcomes.

The chart on page 68 is only an example. If students are familiar with probability, have them make probability estimates for the outcomes.

Part B

1. Conduct a general classroom discussion on the decision charts. What facts turned out to be most pertinent to The Decision? What are some value-based issues which science cannot resolve?

2. Divide students into groups and have them use their decision charts to weigh the options using importance bars. Have students decide as a group.

3. Class discussion: In this activity, students make a decision as a town council member. What do the students personally think about speed limits? If each student voted in a town referendum on the speed limit, how would they vote?

Questions for discussion: How would The Decision change if the road were in a residential street with many young children? What is the best way for the government to improve driving safety?

4. For homework, have students write a few paragraphs in their journals about what they learned.

Extension

Air bags are a controversial topic in the news. Air bags do protect most adults from injury in crashes. New cars in the United States are required to have them installed. However, small children and small people in the front passenger seat have been injured and even killed by air bags. Students should consider whether air bags should be mandatory in all U.S. passenger vehicles.

> **NOTE**
> *Use Teaching Plan: Guided Activities (page 34) to structure the class work.*

EXPERIMENT

Newton's Second Law of Motion states that the acceleration of an object is directly proportional to the net force exerted and inversely proportional to the objects' mass. Experiment with small carts or toy cars loaded with different masses. Roll them down a ramp and measure how much "bashing" force they exert on folded paper tents or other small objects. Then, roll cars of the same mass down ramps of different angles. Have students apply what they've learned to motor vehicle collisions.

	Raise speed limit	Don't raise speed limit
Reduce accidents and fatalities	*Accidents that occur are likely to be more severe.* *Police could be freed from speed traps to concentrate on reducing drunk driving and general policing duties.*	*Speed limits may or may not affect total accidents. Accidents may have more to do with road design, traffic signs, and alcohol involvement.*
Reduce costs to town	*Costs may increase if accidents increase, but it is not clear that they will.*	*Costly accidents may still occur for other reasons.* *Police will be needed to enforce speed limit which take them from other policing duties.*
Reduce emissions	*Emissions are higher at increased speeds, but many drivers are already driving at the higher speed.*	*If the lower speed is strictly enforced, emissions will be slightly lower.*
Increase revenues for town	*Town will lose revenues from speeding tickets if speed limit is raised.* *Accidents contribute business to local car repair shops and hospitals.*	*You can collect revenues from speeding tickets.*

SAMPLE DECISION
CHART: Speed Limits

Assessment

Review the students' journals using Appendix D. Assign the Extension, or have students choose and research their own public safety issue.

Resources

Gartrell, Jr., Jack E. 1989. *Methods of Motion: An Introduction to Mechanics.* Arlington, VA: National Science Teachers Association.

Internet sites

National Highway Traffic Safety Administration
http://www.nhtsa.dot.gov/
Many facts and figures on safety, economics, and prevention of vehicle accidents. Balance this information with that at the NMA site (see below).

National Motorists Association
http://www.motorists.com/
Advocate group, but a good resource for the anti-speed limit argument.

Roller Coasters

STUDENT
SECTION

Background

Amusement parks are great places to study physics. They are full of motion in all different directions and much of it at high speed. Twisting, turning, spinning, upside-down motion, and high speed dips give a thrill that many people line up to experience. The rides in amusement parks generally fall into categories of spinning rides, roller coasters, and miscellaneous such as bumper cars, pendulum rides, etc. Amusement park rides use physics to take riders to what seem to be outrageously risky physical limits.

This activity focuses on the biggest amusement park attraction—the roller coaster. There are several features that make them exciting. Non-looping sections of the roller coasters combine speed with cornering. Many go through an elaborate system of twists and turns, hills and valleys. Most modern roller coasters have a loop in which the roller coaster cars turn upside down at the top of the loop. Many roller coasters have a free fall portion of the ride, in which the cars seem to drop several stories straight down very rapidly. Some roller coasters plunge into water, splashing it in front of the cars and onto the passengers. Let's explore three sources of information for making a decision about riding a roller coaster. First we'll look at the physics and engineering of roller coasters, then we'll look at safety statistics, then we'll read about some roller coaster incidents.

Roller Coaster Schematic

Key:

G- = miminum potential energy

G+ = maximum potential energy

K- = minimum kinetic energy

K+ = maximum kinetic energy

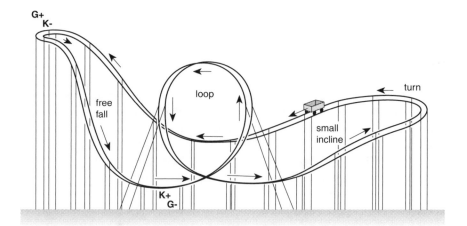

Safety statistics

♦ The chances of being killed in an automobile accident are 42 in one million.

♦ The chances of being struck by lightning are eight in 100 million.

♦ The chances of dying of a cold are six in one million.

♦ Based on riding 100 amusement park rides a year, the risk of death on a ride is four in 100 million.

♦ Each year, people in the U.S. incur: 13,820 injuries on exercise equipment; 11,046 soccer injuries; 6,101 horseback riding injuries; and 5,571 injuries playing billiards.

♦ The U.S. Consumer Product Safety Commission states that since 1973, almost 5,500 persons were treated in hospital emergency rooms for injuries associated with amusement rides.

♦ When you include the 225 to 300 million people who ride carnival rides, amusement rides are ranked 164th on a list of 175 activities in terms of the number of product-related injuries.

Roller coaster incidents

On May 25, 1996, a wheel fell off a Coney Island roller coaster. Two riders were slightly injured. A metal retaining clip that holds the bolts fastening the rear wheel snapped. The bolts were loosened and one wheel fell off. The car immediately skidded onto its belly and jerked to an abrupt stop. One rider banged her head into a support post causing a cut and heavy bleeding. Since 1990, there have been 15 accidents on all amusement rides in New York City. But on the Cyclone, a roller coaster in a New York City amusement park, there have been more than 20 million rides without a single accident.

In England, in 1972, a packed roller coaster slid backward down a slope and broke through wooden barriers. The runaway train resulted in the deaths of five children. Twelve others were injured. The accident was on a wooden roller coaster. Most wooden roller coasters have been replaced by safer, high-strength metal versions with high-tech safety equipment.

The Decision

You are going to an amusement park with your friends. Should you ride the roller coaster? Separately, consider the schematic of the roller coaster, the safety statistics, then the reading about roller coaster incidents. What would your decision be based on each? What would your decision be based on all three?

Roller Coasters

Student Handouts

Student section (pages 69–70)
Appendix B: Summary of Decision Making
Appendix C: Decision Chart

Objectives

Students will:

♦ Apply their knowledge of physics and engineering to roller coasters, and consider the design of safety features.

♦ Analyze levels of risk of riding a roller coaster and other activities.

♦ Develop a decision chart on whether or not to ride roller coasters.

Terms for Understanding

Momentum, vector, velocity, Newton's laws of motion, potential energy, kinetic energy

Procedure

Part A

1. Have students read "Roller Coasters" (student section). After a brief discussion of the topic, students should write a preliminary decision in their journals.

2. Consider the schematic of a roller coaster on the student page. There are labels for maximum and minimum kinetic energy. Have students use their knowledge of Newton's laws to identify additional places where kinetic energy—and therefore stress on the roller coaster structure—would be high or low. You may want to begin this process by asking students if they can remember the bodily sensations they felt at various stages of a roller coaster ride. Make a chart of this information, then correlate it to stress on the roller coaster structure.

3. Now consider the ways engineers might build roller coasters to handle these kinds of stress. This is an opportunity for groups of students to experiment with models, using model railroad track with various inclines, banking of the curves, and masses of cars. Engineers design roller coasters based on these forces. They determine maximum loads, then add a safety margin. Students

can work with models to determine "failure rates," then consider what safety margin they would consider acceptable in the inclines and banks of their models.

Engineers also select materials which can withstand the repetitive loads of roller coaster cars. Consider the flexing ability—and failure rates—of different types of wood, plastic, or metal to simulate this factor. Engineers will plan for maintenance of cars and roller coaster structure, but whether maintenance is carried out will be controlled by the user.

4. Ask students whether or not they would consider it safe to ride a roller coaster based on their perception of the engineering goals and challenges of roller coasters. For sophisticated students, you may want to introduce topics that influence the actions of roller coaster operators: liability for injuries, insurance regulations, and inspection regulations. As a class, identify preliminary goals and options for The Decision. Students should keep notes during this discussion. In groups, or as homework, students should create a decision chart with goals, options, and outcomes. The chart on page 72 is only an example.

Part B

1. Consider the statistics in the Student Section. Have students base a new decision chart on their perceptions of the statistics, combined with their experience with design engineering in Part A. Does this change their perceptions? Have a short class discussion on using different forms of information in decision making. If students are familiar with probability, have them make probability estimates for the outcomes.

2. Divide students into groups and have them use their decision charts to weigh the options using importance bars. On a scale of -100 to 100, have students value the outcomes. Students should consider and discuss the most important and least important outcomes. For example, "Being safe" would have a value of 100. "Death" might have a value of -100. "Socializing with friends" would have a less significant value of 5. Similarly, "enjoying yourself" might have a value of 20.

> **NOTE**
> *Use Teaching Plan: Guided Activities (page 34) to structure the class work.*

3. Students can now compute expected values for death by multiplying the probability of death by the relative value. We know that the probability of death from riding the roller coaster is 4 in 100 million. This is equivalent to .00000004. Multiplying this number by a relative value of -100, the expected value for death while riding a roller coaster would be -.0000004. The chance of death while riding a roller coaster is zero if you do not ride one. The probability of outcomes, except for avoiding death, is 100 percent (or just 1).

Pose the following questions to your students: Since the risk of death is so small, is it really worth considering? And if it is worth considering, then how much weight should one put on it?

Part C

1. Consider "Roller Coaster Incidents" in the Student Section. Have students base a new decision chart on their perceptions of the narrative, combined with their experience with design engineering in Part A and statistics in Part B. Does the narrative change their perceptions?

2. Ask students to analyze their final decision. How much of it was based on how much they enjoy riding roller coasters? What was the role of personal preference, or bias, in making their decision? What was the role of their own experimentation with how roller coasters work in making their decision? Have students explain their answers.

3. For homework, have individual students write a brief (one or two paragraph) essay in their journals about what they learned.

Extension

Decision charts can be developed based on the other data provided on deaths from lightning, colds, and automobile crashes.

Assessment

Review the students' journals using Appendix D. Assign the Extension to evaluate student ability to apply learning, or have students choose and research another related issue.

Resources

Unterman, Nathan A. 1990. *Amusement Park Physics.* Portland, Maine: J. Weston Walch.

Internet sites

Roller Coaster FAQ
http://www.faq.rollercoaster.com/

The Physics of Roller Coasters
http://www.glenbrook.k12.il.us/gbssci/phys/projects/yep/coasters.html/
Links to many roller coaster sites.

SAMPLE DECISION
CHART: Riding a
Roller Coaster

	Ride the roller coaster	Do not ride the roller coaster
Enjoy yourself	*If you enjoy roller coasters, then enjoyment will be very high; if you do not like roller coasters then enjoyment will be low.*	*You will have no enjoyment from roller coaster.*
Be safe	*Death or injury is very unlikely, but still a possible outcome.*	*You will avoid death or injury from a roller coaster accident.*
Socialize	*It's fun to ride roller coasters with a group.*	*You will be left out if your friends ride the roller coaster and you do not.*

Recycling

Background

Imagine you live in Grant County, an urban community with 120,000 residents. The community has no recycling program; it collects all of its garbage at curbside and hauls it to a landfill. However, the cost of disposing of waste in the landfill is rising as the facility begins to fill up. Community opposition is making it more difficult to build new landfills in areas. There is also growing public support for recycling as an alternative to throwing everything away; proponents say it will save forests and other natural resources. The U.S. Environmental Protection Agency has established a national goal for recycling 25 percent of the municipal waste stream. Many communities throughout the nation have already established recycling programs; in most cases, they reach or even exceed the national goal within the first three years of operation.

The Decision

Grant County has decided to investigate whether to start a mandatory recycling program. The county administrator appoints you to a committee to examine the costs, benefits, and tradeoffs of starting this program. The county has stipulated that any program be a mandatory, curbside collection effort done by a private company, not municipal employees. There are three major factors to evaluate in the decision: how much the county can recycle, what it will cost, and the environmental impacts of the program. Will your committee recommend that Grant County begin a recycling program?

Criterion A: How much can the county recycle?

Grant County's current waste stream for next year is estimated to total 87,600 tons. The total waste stream includes all trash generated by private residences, commercial organizations, industries, and institutions. Your program would serve the 120,000 residents, who generate 50 percent—or *43,800 tons*—of all garbage in Grant County.

The total cost of garbage collection service is broken down into three main categories: the fee paid to the private hauler to collect the trash; a "tipping fee" paid to landfill operators who accept the garbage; and an administrative cost for the county to oversee the collection. The county's annual budget for trash collection is shown in Chart 1.

Worksheet 1 shows the average breakdown of municipal solid waste for the United States in 1994. The figures include waste generated by residential, commercial, institutional, and industrial sources.

Expense Category	Cost	Garbage Collected (in tons)	Total Costs
Garbage Collection	*$55 per ton*	*43,800*	*$2,409,000*
Tipping (Disposal) Fee	*$45 per ton*	*43,800*	*$1,971,000*
Administrative Costs	*$2,500,000*	——	*$2,500,000*
Total Cost			***$6,880,000***

CHART 1:
Annual Grant County
Garbage Collection Costs

Questions

1. Fill in Worksheet 1: Using Grant County totals, compute the amount of each type of garbage that Grant County's residences produce on average.

   Assume you will be able to meet the national averages for recycling each type of material. Use information from Worksheet 1 to answer the following questions.

2. On Worksheet 2, calculate the total annual amount of recyclables that could be collected in Grant County based on Worksheet 1 and national percentages of recycled materials.

3. Does this total allow the community to meet the recommended 25 percent recycling rate?

4. What actions could Grant County take to increase recycling? What categories hold promise? What are the potential costs?

WORKSHEET 1:
Average Solid Waste Production, U.S. and Grant County

Waste Material	Percent of Total Waste (National Average)	Grant County Totals (tons)
Newspaper	6.5	
Other Paper*	32.4	
Yard Trimmings	14.6	
Plastic Beverage Containers**	.6	
Other Plastics	8.9	
Aluminum & Steel Containers^	2.5	
Other Metals	5.1	
Wood	7.0	
Food Waste	6.7	
Glass	6.3	
Other	9.4	
Totals	**100**	**43,800**

* Includes magazines, corrugated cardboard, office paper, milk cartons, and other paper products
** Includes soft drink, milk, and water containers
^ Includes primarily drink and food cans

WORKSHEET 2:
The amount of recyclables that can be collected in Grant County

Waste Material	Grant County Totals (tons)	Percent Recycled (Average)	Total Recycled (tons)
Newspaper		52.9	
Plastic Beverage Containers		45.9	
Glass		25.8	
Aluminum & Steel Cans		40.1	
Totals			

Criterion B: Recycling cost factors

Now that we know how much and what kinds of materials will be recycled, the committee needs to estimate costs. There are four cost components.

Collection fee

This is the cost of curbside collection. The typical collection method in the United States is to have separate trucks collect the recyclable materials. The estimated collection fee is $50 per ton.

Administrative costs

The overhead expenses for the recycling program are divided into two areas: start-up costs and on-going oversight expenses. The start-up expenses include costs associated with establishing and staffing a recycling office, bidding the contract, and producing recycling containers for distribution to residents. Administrative costs include operating the recycling office, monitoring contracts, inspections, enforcement, public relations, and community outreach work. The estimated administrative costs are: start-up, $3 million; continuous, $2 million.

Cost avoidance

The county will save money by diverting trash from the landfill. For Grant County, this amounts to approximately $100 per ton ($55 per ton for collection and $45 per ton in tipping fees at the landfill). The county also will save money in administrative fees. The estimated cost savings is the total number of tons recycled multiplied by $100 per ton.

Recycling payment

The recycling payment is the money the government will pay to or receive from a recycling company. If demand for the material is high, the county will receive money from recyclers; if demand is low, the government might have to pay to have the material taken away. For example, when New York and other large cities began mandatory recycling programs, the price for recycled newspapers and other goods dropped sharply. Prices eventually rose as the supply stabilized and recyclers added new capacity for handling the materials.

Estimating recycling payments is difficult. Your committee decided to take a five-year average cost to estimate. See Chart 2 for results.

Waste Material	1992	1993	1994	1995	1996	Average
Newspaper	$150	$50	$65	($20)	($30)	$53
Glass	($100)	($30)	$45	$10	$15	($60)
Aluminum & Steel	($25)	$20	$10	$40	$5	$50
Plastics	$50	($15)	($50)	$10	$30	$25
Average	**$75**	**($15)**	**($20)**	**($40)**	**$20**	**$68**

CHART 2: Payment Per Ton for Recyclable Materials* (Dollars per ton)

*Payments made by communities to recycler to accept materials are shown in parentheses.

Questions

1. Fill in Worksheet 3 to calculate the estimated average payments for Grant County recyclables. Given the substantial cost fluctuations, use the averages from Chart 2.

2. Fill in Worksheet 4 to calculate how much recycling will cost Grant County.

3. How would these cost figures change in the second year of the program?

4. What other savings might exist that are not accounted for in these calculations?

5. Calculations for Question 2 are based on the assumption that mills are paying $53 per ton for newspapers. However, as several years go by, the market becomes saturated as more communities begin recycling programs. In general, what will be the impact on Grant County's program if mills require the county to *pay* $50 per ton to accept the newspapers? Will the price of recycling rise or fall?

6. As land for garbage disposal facilities becomes scarcer, the "tipping fee" often rises substantially. Assume that the tipping fee increases from $55 per ton to $100 per ton. In general, what impact does this have on overall costs? Will the cost of recycling rise or fall?

7. What limitations exist in the calculation methods you have used?

WORKSHEET 3:
Estimated Average Payments
for Grant County
Recyclables

Waste Material	Average Payment (cost)	Amount Recycled (tons)	Total Payment (expense)
Newspaper	$53		
Glass	($60)		
Aluminum & Steel	$50		
Plastics	$25		
Totals	**$68**		

WORKSHEET 4:
Recycling Expenses
for Grant County

	Expense
Collection Fee	
Program Startup	
Administrative	
Cost Subtotal	
Recycling Revenues	
Cost Savings	
Total Cost	

Criterion C: Environmental impacts of community recycling

Our analysis of recycling has concentrated primarily on the economic impact of running the program. Recycling also has a substantial environmental impact that must be considered when making any decision. Recycling can preserve raw materials, such as minerals and trees. For example, the United States recycled 27.1 million metric tons of paper, paperboard and wood in 1994, approximately 31.4 percent of the 86.3 million tons of paper waste generated that year. It is estimated that a family of four uses approximately 1.36 metric tons of wood and paper products per year, which requires four trees 45.7 centimeters in diameter and 30.5 meters tall. This amounts to approximately three trees used for each metric ton.

Forests serve many valuable purposes. They provide habitat for the vast array of animal and plant species required to maintain diverse food webs. Forest ecosystems act as the planet's lungs, absorbing atmospheric carbon dioxide and producing oxygen and complex carbohydrates. The forests also provide aesthetic and recreational benefits to humans.

According to one estimate, the ecological value of a typical tree is worth almost $200,000 in oxygen production, air and water cleansing, habitat provisions, soil fertilization, erosion control, and other benefits. By contrast, the same tree sold as timber on the commercial market might fetch $600.

Questions

1. Suppose that Grant County expanded its paper recycling program to recycle 3,300 tons of newspaper, wood, office paper, cardboard, and paperboard. What would be the potential financial impact for the nation as a whole?

2. How does this calculation affect the costs of Grant County's recycling program?

TEACHER SECTION

Recycling

Student Handouts

Student section (pages 73–77)
Appendix B: Summary of Decision Making
Appendix C: Decision Chart

Objectives

Students will:

- ◆ Evaluate how much a fictional county can recycle. They will estimate how much and what type of municipal solid waste is generated in the county each year.
- ◆ Estimate costs. The students will fill out worksheets and answers questions concerning the costs of recycling.
- ◆ Learn about environmental benefits and drawbacks to recycling.
- ◆ Make a decision about whether a recycling program should be started in a fictional county.

Terms for Understanding

Tipping fee, cost avoidance, landfill, municipal waste

Procedure

Part A

1. Have students read "Recycling" (student section). They should not do the worksheets and calculations yet. After a brief discussion of the topic, have the students write a preliminary decision in their journals.

2. Discuss The Decision. Whose decision is it? Discuss the difference between individual, local, and national decisions to implement recycling programs.

3. Risk assessment. This activity includes detailed information derived from the groups listed in Resources. An introductory discussion will review the information presented to students. Then, for homework, students will fill out the worksheets and make cost calculations.

NOTE

Use Teaching Plan: Guided Activities (page 34) to structure the class work.

Evaluate how much the county can recycle.

Have students look at Worksheet 1 on page 74 (student handout). Ask your students:

What type of waste makes up the largest percentage of the total? According to the Environmental Protection Agency, the United States' total municipal solid waste generation for 1994 was 188.2 million metric tons. The breakdown by type is given as percentages of the total in Worksheet 1. Students may be surprised that yard trimmings (primarily grass cuttings and branches) make up such a large percentage of municipal solid waste. Did they know that plastic beverage containers make up such a small percentage of the waste stream?

Which materials commonly used in the home can be easily recycled? Commonly recycled materials include newspapers, mixed glass (brown, green, and clear), metal cans (primarily aluminum), and many types of commonly-used plastic containers (primarily milk jugs and soda bottles). These products can be easily broken down into raw materials for use in making new products. Some communities have separate programs for recycling grass clippings and brush.

For the purposes of this activity, encourage students to start with the basics—glass, metal cans, plastics, and newspapers. Students will probably be familiar with these materials as being part of most recycling programs. Limiting the number of variables also will simplify the calculations.

Estimate costs.

Have students read page 75 on recycling cost factors. Make sure they understand the four factors.

Review Chart 2 (page 75) with them; explain that this chart shows how demand for certain materials varies over time. Sometimes recyclers pay money for materials, and sometimes there is an oversupply. Then, the county will have to pay the recycler to take the materials.

Environmental impacts of recycling.

Have your students read page 77. Ask them the following questions:

What are some of the environmental drawbacks of recycling? Recycling is not free of pollution. Trucks are used to collect the recyclables and to transport them to centers. The recycling centers also produce pollution by using energy. Chemicals used in the recycling process must be carefully handled to avoid spills that contaminate rivers, groundwater, and soil. These offset the environmental benefits of not harvesting trees or extracting minerals from the ground.

What are some possible shortcomings to estimating environmental values? The financial estimate for the worth of a typical tree is subjective. One can place a value on what trees do—water cleansing, habitat provisions, and soil fertilization—and the negative impacts of removing trees, such as soil erosion. There is also financial and environmental impact of replacing trees through replanting. By contrast, it is much easier to estimate the cost of logging a tree and what it can be sold for on the commercial market.

This estimate also misses the financial impact of the logging industry. Recycling can limit the growth of the multi-billion dollar forestry industry, having an effect on workers in the field and the economy in general. On the other hand, the recycling industry also creates jobs and provides other benefits to the economy.

4. Homework: Have students complete the worksheets and answer the questions on pages 74, 76, and 77.

Part B

1. Review the answers to the questions assigned as homework. Then, return the discussion to The Decision. What should the committee recommend? Use the sample decision chart below as a guide.

2. Break students up into groups and have them make their own chart to weigh the options using importance bars. Have students make a group decision.

3. Lead a class discussion of the group decision. The discussion should include questions such as: What option did they pick? How does the final decision compare to the preliminary decision they made the first day? What uncertainties in the outcomes make the decision difficult?

4. For homework, have individual students write a brief (one or two paragraph) essay in their journals about what they learned.

SAMPLE DECISION CHART: Recycling

	Begin recycling program	Do not begin recycling program
Save money	*Recycling programs can cost communities millions of dollars to implement and operate. Costs can fluctuate substantially depending upon the price recyclers will pay for materials.*	*Costs for landfill disposal can rise when land becomes scarcer. There are costs to closing and monitoring existing landfills and opening new ones.*
Save land & trees	*Recycling programs can save land by reducing the amount of trees used and minerals mined. Recycling also extends the life of existing landfills and limits the size of new ones.*	*Modern forestry practices allow for the regrowth of forests. There are plenty of forests in the country and land for garbage dumps. However, heavily populated areas often lack land for garbage dumps.*
Preserve the environment	*Preserving forests, reducing mining, and limiting landfill can have positive impacts on the environment.*	*Recycling programs create pollution through the collection, transportation, and reprocessing of the materials. Land also must be used to site trash transfer stations and recycling centers.*

CRITERION A:
Worksheet 1
Answers

Waste Material	Percent of Total Waste (national average)	Grant County Totals (tons)
Newspaper	6.5	2,847
Other Paper*	32.4	14,191
Yard Trimmings	14.6	6,395
Plastic Beverage Containers**	.6	263
Other Plastics	8.9	3,898
Aluminum & Steel Containers^	2.5	1,095
Other Metals	5.1	2,234
Wood	7.0	3,066
Food Waste	6.7	2,935
Glass	6.3	2,758
Other	9.4	4,178
Totals	**100**	**43,800**

* Includes magazines, corrugated cardboard, office paper, milk cartons, and other paper products
** Includes soft drink, milk, and water containers
^ Includes primarily drink and food cans

CRITERION A:
Worksheet 2
Answers

Waste Material	Grant County Totals (tons)	Percent Recycled (average)	Total Recycled (tons)
Newspaper	2,847	45.9	1,307
Plastic Beverage Containers	263	40.1	105
Glass	2,758	25.8	712
Aluminum & Steel Cans	1,095	52.9	579
Totals	**6,963**	**38.8**	**2,703**

CRITERION B:
Worksheet 3
Answers

Waste Material	Average Payment (cost)	Amount Recycled (tons)	Total Payment (expense)
Newspaper	$53	1,307	$69,271
Glass	($60)	712	($42,720)
Aluminum & Steel	$50	579	$28,950
Plastics	$25	105	$2,625
Totals	**$68**	**6,963**	**$58,126**

Answers to student questions

Criterion A: How much can the county recycle?

1. Answers to Worksheet 1 are found on p. 80.

2. Students should multiply the amount of material Grant County produces each year (column 3 in Worksheet 1) by the national recovery average for the relevant category (column 2 in Worksheet 2). For answers, see p. 80.

3. No. The 2,703 tons is only 6.2 percent of the county's residential waste stream. Stress to the students that while the rate of recovery for each category is high, the materials being collected only make up 15.9 percent of the total waste stream.

4. One step is to try to increase the participation rate among residents. Additionally, residents might be interested in finding out ways to produce less waste. This can be done through public education and public relations efforts. The specific costs of these efforts, which will come out of the administrative budget, are difficult to gauge accurately.

Another solution is to broaden the types of materials collected under each category. For example, Grant County might include magazines, corrugated boxes, office paper, telephone directories, and additional types of plastics. Other potential recyclable items include tires, scrap metal, and automobile batteries.

Yard trimmings make up 14.6 percent of the waste stream; many communities have composting operations in which grass, leaves, and brush are turned into mulch for gardens. To begin a composting operation, the county might have to set aside land and pay additional collection and administrative fees. However, the county would save money by avoiding tipping fees at the landfill.

Criterion B: Recycling cost factors

1. Answers to Worksheet 3 are on page 80.

2. Recycling expenses for Grant County are listed above.

3. Costs would decrease because all the expenses for program startup are borne during the first year. Minus the $3 million cost for startup, the county's cost would be $1.8 million.

	Expense
Collection Fee	$135,150
Program Startup	$3,000,000
Administrative	$2,000,000
Cost Subtotal	**$5,135,150**
Recycling Revenues	$58,126
Cost Savings	$270,300
Total Cost	**$4,806,724**

CRITERION B: Worksheet 4 Answers

4. The county could save money in several areas. For one, the life of the landfill now used for trash disposal could be extended for many years. Closing a landfill can be expensive because owners must meet extensive environmental standards and conduct substantial monitoring to prevent ground-water contamination. Another area of savings involves lowered expenses for collecting regular garbage. And with less trash to collect, the county might consolidate existing routes.

5. This change would increase Grant County's costs because the county would be paying more money to haul away newspapers.

6. The county would save more money because recycling helps the county avoid tipping fees.

7. The activity is limited in that it assumes that the number of residents and the amount of garbage generated will stay equal during the first several years of the program. This may not occur in a real county, where populations grow as new housing is built and families expand. Some large American cities have actually lost population over a period of time as people moved to suburban regions.

Criterion C: Environmental impacts of community recycling

1. At a rate of three trees per ton, recycling 3,300 tons would provide raw material equivalent to 9,900 trees. The net benefit of that at $200,000 per tree would be approximately $1.98 billion in benefits to the nation as a whole.

2. It doesn't. Although the society as a whole benefits, Grant County does not receive any funds to offset its financial costs.

Extension

Many communities that have recycling programs are facing a problem of just how far to extend them. Ask students to consider what effort would be required to expand Grant County's recycling rate from 25 percent to 50 percent. What additional steps would be required? How much might that cost? Would residents be willing to participate if it meant spending additional time sorting their garbage?

Assessment

Review the homework questions.
Assign the Extension to evaluate student ability to apply learning; or have students choose and research another recycling issue.

Resources

Costanza, Robert, et al. 1997. "The Value of the World's Ecosystem Services and Natural Capital." *Nature* 387 (15 May).

American Forest and Paper Association
1111 19th Street, NW, Suite 800
Washington, DC 20036
Phone: (800) 878-8878
Internet site: http://www.afandpa.org/
Provides information about paper manufacturing, forest management, and recycling.

National Recycling Coalition
1727 King Street, Suite 105
Alexandria, VA 22314
Phone: (703) 683-9025
Provides information and contacts about recycling issues.

Internet sites

Recycler's World
http://www.recycle.net/
A worldwide trading site for information related to secondary or recyclable commodities, by-products, used & surplus items or materials.

Solid Waste Association of North America
http://www.swana.org/
Email: swana@millkern.com
Provides information on issues related to solid waste.

U.S. Environmental Protection Agency
http://www.epa.gov/
Main EPA site.
EPA Office of Solid Waste and Emergency Response:
Non-Hazardous Waste Page
http://www.epa.gov/epaoswer/non-hw/index.htm
Provides information about solid waste and recycling programs. The office publishes an annual report, Characterization of Municipal Solid Waste in the United States, *that provides a complete statistical breakdown of U.S. waste generation.*

A Local Decision

TEACHER SECTION

Student Handouts

There are no student pages for this activity. Find local news articles or publications of the decision-making body.

Appendix B: Summary of Decision Making

Appendix C: Decision Chart

Objectives

Students will:

♦ Research a recent decision made by a local decision-making body.

♦ Analyze how the decision was made and compare the local decision-making methods to methods presented in this book.

Background

Every city, town, village, or school has boards that make decisions for the rest of the people. Common decisions that affect towns could be on what to build and where, or if new industries will affect environmental or economic systems. Among other duties, school boards decide how to structure classes, or what books to use in the schools.

Students can learn more about decision making by observing how local legislative bodies deliberate on issues and reach decisions. This process also shows students how local decisions can impact their lives, sometimes even more profoundly than national decisions at the congressional level.

Students can also formulate their own opinions about local issues by using the decision-making skills learned in this book. This lesson helps connect the decision-making activities with those skills, and the abstract issues that they have been studying, to real issues in their local community.

Procedure

1. Have students research or provide them with a list of up to five decision-making entities in your local area. Examples can include: the local school board, the village or town council, the committees of that council dealing with specific issues, or the chamber of commerce. Each community will have its own kinds and structures. Explain what each organization does and why it exists, and what kind of decisions it usually makes.

Other considerations could be:

♦ How does the committee gather its information? Does it research an issue, or rely on outside experts or testimony from concerned citizens?

♦ How does the committee develop options for the decision?

♦ Once the decision is made, is there a procedure to reevaluate it when further evidence comes in?

♦ How does the committee gather scientific evidence if the issue involves science in some way?

♦ How "expert" is the committee in making decisions? Do any members have formal education in decision making or in the subject being debated?

2. Pick one or two of these bodies and find out the most recent of their decisions. Try to find an issue as close to your curriculum as possible. As a class, develop a decision chart with the students to show different options that the committee considered and what they decided.

3. For homework, have the students develop their own charts and make their own decision. Use the Summary of Decision Making (page 119) as a guide. Discuss their conclusions.

4. Ask the students how they could contribute to this process. Does the public have an opportunity to give input during the decision process? Depending on the committee or the issue, it may be possible to have a committee member visit your class to discuss the decision they made and how they arrived at it.

Extension

Students could form a mock committee and debate the decision. Then have the class or the committee members vote. Was the class decision the same as the real committee's? Why or why not?

Assessment

Review the students' journals using Appendix D. Have students choose and research another local issue.

Resources

Local newspapers usually carry summaries of recent decisions by governmental bodies. Formal community groups may also publish newsletters.

Internet sites

Note: These sites may be useful in researching background on local issues.

Congress.org
http://congress.org/
A guide to Congress, including a list of committees and how to find your member.

StateLaw
http://lawlib.wuacc.edu/washlaw/uslaw/statelaw.html
State government and legislation information.

New York State Government
Information Locator Service
http://www.nysl.nysed.gov/ils/
This is just an example of a state government information service. Look for your state's site.

PART THREE

Independent

Exercises

TEACHING PLAN:
Independent Exercises

Objective

Each exercise is a brief introduction to an issue that students can explore independently.

Required Student Skills

Students should be familiar with the Part One concepts and have completed one guided activity before doing the independent exercises.

Time Management

The exercises can be introduced in one 50-minute class period. However, most of the student work should be done outside of class. A short wrap-up discussion will be useful. Note: In the activities "Asbestos" and "Old-Growth Forests," students role-play different constituencies in the decision. See student sections for directions.

General Procedure

First session

1. Before beginning a new exercise, photocopy and hand out three items: the student section for the exercise (there are no teacher pages), Appendix B: Summary of Decision Making, and Appendix C: Decision Chart.

2. Have students read the handout. (5 minutes)

3. Conduct a class discussion of the topic under consideration. Make sure students understand the issue. (5 minutes)

4. Have students write a brief, preliminary decision in their journals. Encourage them to use what they know of the decision-making process. (5 minutes)

5. Continue the class discussion on the issue. (25 minutes)

   ♦ Ask the question: Whose decision is it? See teaching box (page 6).
   ♦ Research and discuss a risk assessment for the hazard. Note: An additional day could be spent

on research and discussion to fully define the hazard. See pages 6–9.
   ♦ Make a preliminary list of goals for the decision. See teaching box (page 9).
   ♦ List some possible options (these need not be final). See teaching box (page 11).
   ♦ Elicit ideas about what kinds of data and additional information students might need to make a more informed decision. You may want to keep a running list of student ideas on the chalkboard.

6. After this discussion, students should work on their own. Following the steps in the Summary of Decision Making, they should research the issue, create decision charts, and make a decision. The student section suggests resources and Internet sites for students to explore. They should carefully document their thought processes and research in their journals. Explain to students that you will be using the journals to assess their work.

Second session

Conduct the second session several days later to allow students to do research on their own.

1. Lead a general class discussion of the group decisions. The discussion should include questions such as:

   ♦ What option did you pick? Is the final decision different than the preliminary decision? Why or why not?
   ♦ What uncertainties in the outcomes make the decision difficult?
   ♦ What single most important fact is needed to make the decision more definite?

Assessment

Assess student work using their journals and their participation in class discussions. See Appendix D for more information.

Bovine Growth Hormone

Background

When scientific discoveries are paired with traditional farming techniques, the results can improve both quality and output. Such is the case with bovine growth hormone (BGH). A protein hormone, BGH is produced naturally in the pituitary gland of cattle. Biotechnology has enabled scientists to produce a form of BGH known as recombinant bovine Somatotropin (rbST). Researchers have discovered that supplementing cows' natural levels of this protein improves their efficiency as milk producers. It enables the mammary glands of dairy cows to take in more nutrients from the bloodstream and produce more milk.

Most scientists agree that bST, whether recombinant or natural, has no effect on humans. In fact, all milk contains small levels of the natural form of this hormone. Any trace amounts of rbST are inactive, because research has shown that the bovine forms of this hormone cannot be processed by humans. There is no way to tell the difference between regular milk and milk from supplemented cows.

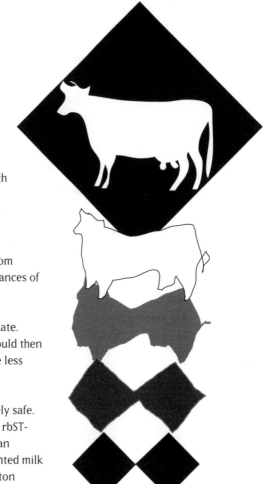

Groups that favor organic farming argue that the U.S. Food and Drug Administration approved rbST without testing the hormone's long-term health effects on consumers. Opponents fear that cows receiving rbST are likely to have higher incidences of udder disease and therefore would be treated with antibiotics. If antibiotic traces remain in milk, the residue can spur allergic reactions and antibiotic resistance in humans. (The National Milk Producers Federation says that safeguards in the milk inspection system prevent this from happening.) A second concern is that using rbST may increase consumers' chances of getting breast cancer.

Other concerns about this hormone focus on economic problems it might create. Some critics worry that rbST will lead farmers to overproduce milk, which would then depress farm milk prices. If this happened, the economic squeeze might drive less efficient dairy farmers out of business.

The FDA reported in 1985 that milk from rbST-supplemented cows is absolutely safe. In 1991, the National Institutes of Health announced that milk and meat from rbST-treated cows is no different from products from untreated cows. The American Medical Association's Council on Scientific Affairs found that rbST-supplemented milk is completely safe; these findings were confirmed by a 1994 study by the Clinton Administration. Vermont passed a mandatory labeling law, and that law was overturned in federal court.

The Decision

You own a chain of dairy farms in the Midwest. Your business has consistently turned a profit, and you have never used hormones or steroids to increase your herd's milk productivity. To take advantage of recent hikes in milk prices, you are considering administering rbST. A local consumer group threatens to boycott your milk, concerned that rbST milk may have unknown effects. Young children, they fear, might be especially affected. After looking into the matter, you find that rbST will probably increase your herd's output by about five percent. And without a boycott, your chain's profits will probably increase about $1 million per year. You don't know how much a boycott will decrease your profits. Is it worth the risk? What do you do?

Resources

U.S. Food and Drug Administration
Office of Consumer Affairs
5600 Fishers Lane
Rockville, MD 20857
Phone: (301) 443-1370
Internet site: http://www.fda.gov/

National Center for Nutrition and Dietetics
American Dietetic Association
216 West Jackson Blvd.
Chicago, IL 60606
Phone: (312) 899-4853

National Milk Producers Federation
1840 Wilson Blvd.
Arlington, VA 22201-3000
Phone: (703) 243-6111

Local resources might include a home economist from your county extension office (cooperative extension system) or a local nutrition professional in your local public health department, hospital, or dietetic association.

Internet sites

Animal Rights Resource Site: Essays
http://envirolink.org/arrs/essays/milk_musings.html
Essay: "Milk, rBGH, and Biotechnology."

Food Marketing Institute
http://www.fmi.org/media/bg/bst.html
Provides answers to frequently-asked questions about rbST, in addition to updates on current FDA and state labeling guidelines.

Chemical Warfare

Background

Most military weapons are based on chemical reactions. Chemical weapons, however, are toxic substances that kill, cripple, sicken, or incapacitate. They first came into widespread use during World War I. The most common types of chemical weapons act on the nervous system. These organo-phosphorus compounds inhibit the transmission of nerve impulses by blocking the action of acetylcholinesterase, an enzyme vital to the transmission of nerve impulses.

Chemical weapons are internationally unpopular because they can impact civilians, especially when chemicals are sprayed into the air. Testing chemical weapons also creates problems. Without testing these weapons on humans, it is difficult to have full knowledge of their effects. Chemical weapons also harm trees, soil, air, and water quality. Currently, only a handful of nations even acknowledge that they develop chemical weapons.

The Decision

Put yourself in the place of a U.S. Senator. It is 1996, and the Chemical Weapons Convention has not been passed. Your state is hoping to receive a large defense contract to build a chemical weapons testing center. Local leaders support the contract because it will bring in jobs. Environmental groups are concerned about the potential health and environmental effects.

CHEMICAL WEAPONS BAN

In this hypothetical exercise, you decide whether to support the use of chemical weapons. But did you know that on April 24, 1997, the U.S. Congress approved the Chemical Weapons Convention? The convention bans the use, stockpiling, transfer, and development of chemical weapons. The international treaty has currently been signed and ratified by more than 70 countries. Opponents argue that the convention is unverifiable—there is no way to tell whether other nations are abiding by the convention's guidelines. The convention's supporters say it will improve safety in both war and peace time, and may help prevent situations such as "Gulf War Syndrome."

What other considerations should go into your decision? Do you want the U.S. to use chemical weapons at all? You may want to research media articles about the use and effects of chemical weapons during the Vietnam and Persian Gulf Wars. What if other nations have chemical weapons and the U.S. doesn't? Might we lose wars? Should you try to get the testing center approved?

Resources

Slesnick, Irwin and Miller, John. 1993. *Real Science, Real Decisions.* Arlington, VA: National Science Teachers Association.
Contains a reading on chemical warfare.

Internet sites

Gulf War Veterans Resource Site
http://www.gulfweb.org/

Stockholm International Peace Research Institute
http://www.sipri.se/projects/chembio.html
Provides details about the 1997 Chemical Weapons Convention.

STUDENT SECTION

Asbestos

Background

Asbestos is a rock that is used for insulation, fire retardants, flooring, shingles, and scores of other purposes. Asbestos is everywhere, including auto-brake pads and building materials. During the past 20 years, this fibrous mineral has been identified as a potential cause of lung disease and cancer. When released into the air, the tiny fibers lodge in the lungs and lead to a multitude of health problems.

Evidence of asbestos-related maladies began in the 1950s when large numbers of workers in mines, shipbuilding, and other asbestos-using businesses contracted lung cancer and asbestosis. As the link between asbestos and health problems became clearer, school officials evaluated walls, floor and ceiling tiles, roofs, and insulation materials that contained asbestos fibers. Across the country, school districts have spent millions of dollars to remove asbestos because they believed that their children were in great danger. The United States has spent about $10 to $30 billion to address asbestos problems.

However, the large-scale asbestos removal from schools and public buildings may have been unnecessary. There is no scientific consensus on how much asbestos in the air is unsafe. And there is now broad consensus among scientists and physicians that asbestos in public buildings is not much of a threat to health. New data have shown that levels of airborne asbestos in buildings with flaking insulation could be as low, or nearly as low, as the air outdoors.

The Decision

Assume that asbestos has recently been found in insulation on your school's heating pipes. Recent data has shown that perhaps asbestos does not pose as serious a health risk as previously thought. However, many people still fear potential health risks. If you have time, do additional research on asbestos and its removal. Look at the health and safety hazards of removal versus non-removal. Your teacher will divide your class into three groups, and assign you one of the following roles: Parents and Teachers, Environmental Protection Agency, or the School Board.

Group 1: Parents and Teachers
This group wants to make sure that the safety of the students is most important.

Group 2: Environmental Protection Agency
This group is following a Congressional request to develop regulations about the safe amount of asbestos.

Group 3: School Board
This group is in charge of allocating funds to schools. Asbestos removal is costly and would necessitate cuts in other school programs. If asbestos does not legally have to be removed, what options do you have?

Develop a list of goals and concerns. When you finish brainstorming, discuss the issue as a class. Then vote on the decision: Should the asbestos be removed from the entire school?

Resources

Cary, Peter. 1995. "The Asbestos Panic Attack." *U.S. News & World Report*, February 20: 61–63.

U.S. Department of Labor
200 Constitution Ave., NW
Washington, DC 20210
Internet site: http://www.dol.gov/

U.S. Environmental Protection Agency (EPA)
EPA Asbestos Ombudsman Clearinghouse/Hotline
Phone: (800) 368-5888
Provides information on the handling, abatement, and management of asbestos in schools, the workplace, and the home. It also provides explanations of recent asbestos legislation.

Internet sites

Asbestos
http://abwam.com/grossing/refasbes.html
This site maintained by Radon Detection Systems to discuss the history, use, and hazards of asbestos.

The Asbestos Institute
http://www.asbestos-institute.ca/
A database with abstracts to more than 9,000 scientific articles, books, and conference proceedings related to the field of fibers and health.

EPA Region 5: Asbestos Standards
http://www.epa.gov/docs/ARD-R5/asbestos/asbestos.htm
Good general information page. See information and guidance documents.

Occupational Safety and Health Administration (OSHA)
http://www.osha.gov/
OSHA maintains regional offices across the country; check local phone listings.

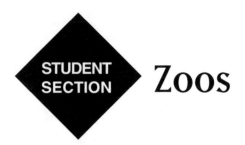

Zoos

Background

People go to zoos to enjoy themselves, learn, and marvel at the wide variety of animals found on Earth. Once simple menageries, many zoos are now changing their mission. During the past 20 to 30 years, zoos have increasingly attempted to create environments similar to animals' native habitats.

The San Diego Zoo was a pioneer in creating natural living environments. The emphasis on natural habitat helps educate the public about the importance of habitat. In conjunction with this new approach, many zoos are taking on the role of preserving and breeding endangered animals. These zoos are viewed as centers of biodiversity preservation.

Zoos are also taking a leadership role to improve environmental education. When children have an opportunity to see endangered species, they are more likely to want to protect them. The Cincinnati Zoo, for example, has developed a broad spectrum of education programs, including lectures, classes, and Internet educational connections. Partnering with Sun Microsystems, Inc., the zoo is setting up interactive multi-media kiosks around the zoo. The kiosks instruct as they entertain.

Some animal rights advocates question whether humans have the right to use animals as resources. Some believe that animals should not be used for food, clothing, or research purposes. They argue that visiting aquariums and zoos is a leisure activity for humans, and that animals should not be used in this way.

A third point of view comes from animal welfare advocates. This group asserts that there are essential uses (such as biomedical research) and nonessential uses (such as entertainment). Animal welfare advocates support animal use practices that can produce widespread benefits to society, but reject nonessential use.

The Decision

An anonymous donor has just given a highly-respected zoo a multibillion dollar donation. The donor's only request is that the funding be used "to the greatest benefit of endangered species and the public." Two proposals have been made. The first proposal is to create a system of satellite zoos around the world. The satellite zoos would be placed in areas that currently have no zoos, and would offer educational programs for the local public. Like most current zoos, the animals would be contained within the parks' boundaries, and caretakers would keep the animals healthy.

The second proposal is to spend the money to preserve natural habitats. The funds would protect the habitat of endangered species. Educational programs would take place outside these habitats so that humans do not interfere with the animals' natural surroundings.

As the Zoo's Board of Directors, your group must decide what goals can and should be achieved by a modern zoo. You are aware that the zoo's goals will have a large influence on future zoos. What do you choose to do with the funds?

Resources

Bostock, Stephen. 1993. *Zoos and Animal Rights: The Ethics of Keeping Animals*. London and New York: Routledge.

Maple, Terry, and Archibald, Erika. 1993. *Zooman: Inside the Zoo Revolution*. Marietta GA: Longstreet Press.

Siebert, Charles. 1993. Fear of Zoos. *The New York Times*, February 9.

Internet sites

Animal Rights Resource Site
http://www.envirolink.org/arrs/
Links to a variety of animal rights organizations; provides an overview of beliefs held by animal rights and animal welfare advocates.

Cincinnati Zoo and Botanical Garden
http://www.cincyzoo.org/
Links to descriptions of the zoo's programs that promote wildlife and conservation education.

Pro Animal Use Site
http://mtd.com/tasty/
Useful site which connects to many husbandry organizations and other groups advocating the use of animals for human needs.

The San Diego Zoo
http://www.sandiegozoo.org/

The Smithsonian Zoological Park (The National Zoo)
http://www.si.edu/natzoo/

Marine Resources

Background

Fishing communities around the world are worried about the future. The fish catch increased rapidly after World War II, but fish supplies leveled off in the late 1980s, and have fallen since 1989. Many fishing communities are finding that they harvest less even if they work harder. Some communities are even being banned from their traditional livelihood. The marine harvest has now fallen in all but two of the world's 17 major marine fishing areas.

Worldwide, fish and other products of the sea account for 16 percent of the animal protein consumed by humans. Yet fish are becoming expensive even in industrial countries, and some once-common species are no longer readily available in supermarkets.

After decades of using bigger boats and more advanced hunting technologies, the oceans are fished nearly to the limits. Analysts from the United Nations Food and Agriculture Organization (FAO) found overfishing in one third of the fisheries they reviewed. The FAO also found depleted fish populations in nearly all coastal waters around the world. In the last few years, more than 100,000 people who fish for a living have lost their jobs, and many more could be out of work in the next decade if this trend continues. But hard times among marine fishing communities can turn fishing towns into ghost towns and cause fishing cultures to disappear. Small scale fishing forms the backbone of community and cultural diversity along many of the world's coasts.

However, a very promising form of fish production is increasingly being used. Aquaculture, also called fish farming, is a method in which fish are grown in pens in natural water bodies. In many cases, large quantities of fish are successfully raised and profits made. While aquaculture techniques take pressure off of marine fish populations, it can be expensive and labor intensive. Also, some argue that farm-raised fish are not as tasty as naturally-harvested fish.

Fishing facts

- ◆ Marine fish populations fluctuate in natural cycles, so setting sustainable harvest figures that hold true every year may not be possible or desirable.
- ◆ Since the mid-1700s, four marine species have been hunted to extinction: Steller's sea cow, the Caribbean monk seal, the Atlantic grey whale, and the sea mink.
- ◆ Overfishing of cod, flounder, haddock, and other fish costs the New England economy approximately $350 million and the loss of 14,000 jobs annually .
- ◆ For every one pound of shrimp caught by trawlers in the Gulf of Mexico, four pounds of fish are discarded and killed.
- ◆ Numbers of factory trawlers operating off Alaska increased by 540 percent between 1986 and 1993.
- ◆ Over 20 percent of the fish sold in many grocery stores are raised via aquaculture.
- ◆ Industrial nations currently import nearly seven times as much fish as do developing nations.
- ◆ International fishing agreements, most of which are sponsored by the United Nations, are still in their earliest stages. Once completed, only those nations that sign an agreement will be bound to honor its provisions.
- ◆ Land-based, human-caused pollution, especially waste disposal and sewage runoff, are a primary cause of depleted marine fish populations.
- ◆ Scientists predict that Earth's human population will double by the year 2025, putting even greater pressure on already depleted marine stocks.

The Decision

You must decide how to address the problem of rapidly depleting marine resources. But before you make your decision, consider who is best suited to make such a decision. Is it a local, state, national, or international decision? Is it a combination of two or more of these? Whatever you decide, develop your decision chart from that viewpoint.

Resources

Norse, Elliot, ed. 1993. *Global Marine Biological Diversity: A Strategy for Bringing Conservation into Decision Making.* Washington, DC: Island Press.

Weber, Peter. 1994. "Safeguarding Oceans." In *State of the World: A Worldwatch Institute Report on Progress Toward a Sustainable Society.* New York: W.W. Norton & Company.

World Resources Institute. 1994. *World Resources, 1994-1995: A Guide to the Global Environment.* New York: Oxford University Press..

Overfishing: Causes and Consequences. Special issues of *The Ecologist* 25 (nos. 2 and 3).
To order from MIT Press:
Phone: (617) 253-2889;
Internet site: http://mitpress.mit.edu/

Internet sites

National Oceanic and Atmospheric Administration (NOAA)
http://www.noaa.gov/
Main NOAA site.
NOAA: National Marine Fisheries Service
http://kingfish.ssp.nmfs.gov/

Overfishing Backgrounder
http://www.lehigh.edu/~kaf3/books/reporting/fishing.html
This link from Lehigh University's homepage provides a good introduction to unsustainable fishing practices and their results.

Background

You peer inside the refrigerator, and a variety of food awaits you. So many choices—soda, juice, lemonade, and milk are just some of the drinks. And what to eat—fruit, cheese, pudding, cold-cuts? But how healthy are these foods? And how can they affect the way you feel?

By grabbing whatever looks best to you at the moment, you make several decisions about what to eat. While these everyday decisions may not seem very important, nutritionists and physicians say these may be the most important choices you make in your life.

What do you enjoy most about eating? Some examples include taste, feeling good about yourself, and so on. Write a list of about 10 things you personally enjoy the most, and label these "benefits." Next, think about some of the negative aspects of eating, such as poorly-prepared food, the cost of eating expensive food, and the time involved in preparing meals. Make a list of 10 such aspects and label them "drawbacks."

In a perfect world, what would be ideal foods? What would be the ideal results of eating? Some simple answers are that food would taste great and not make you gain weight.

After you have run out of ideas, develop a decision chart. What goals would you have for an ideal diet? Because you live in the real world, it may be more difficult to pin down your options. They can be either vague (prepare food at home, eat fast food, etc.) or specific (vegetables, fatty meats, lean meats, etc.). Decide what foods best fit your goals and options.

The Decision

Your homework tonight is to keep track of what you eat. Keep a pen and paper handy throughout the day, and write down everything you eat and drink. Be honest!

Before class meets again, write one to two pages about how your actual diet compares with the results of your decision chart analysis. Does your diet support or contradict what you identified as benefits? What kind of decision-making process do you use to determine your daily diet? How does this process compare with the decision chart process? Do you think you need to evaluate every single food choice? What is the best way to make decisions regarding personal diet?

Resources

U.S. Department of Agriculture and U.S. Department of Health and Human Services. *Dietary Guidelines for Americans*. 1995. Washington, DC: USDA and USHHS.
To order, write:
Food and Nutrition Information Center
Agricultural Research Service
National Agricultural Library, Room 304
10301 Baltimore Ave.
Beltsville, MD 20705-2351
Phone: (301) 504-5719
Also available on-line, see USDA site below.

Internet sites

Center for Science in the Public Interest
http://www.cspinet.org/
This non-profit education and advocacy organization focuses on improving the safety and nutritional quality of our food supply.

USDA Food and Nutrition Information Center
http://www.nal.usda.gov/fnic/
Excellent resources about food and nutrition.

Dietary Guidelines for Americans
http://www.nal.usda.gov/fnic/dga/

U.S. Department of Health and Human Services: Healthfinder
http://www.healthfinder.gov/
Gateway consumer health information website. Connects the user to reliable agencies and organizations that provide health information to the public.

Hearing Loss

Background

Have you ever had trouble hearing your friends after leaving a loud music concert? If so, you probably noticed a ringing in your ears that went away after a short while. What you experienced is tinnitus, one of the most common forms of hearing loss. More than 50 million people in the U.S. experience some form of tinnitus.

For most people, the ringing in their ears ends shortly after the exposure. While frustrating, it has little lasting effect. But tinnitus can be a very serious auditory problem—each year more than 12 million people seek medical advice for persistent tinnitus symptoms.

Tinnitus occurs when the auditory nerve has been shocked. The brain interprets this as noise, which is why you hear ringing. When noise is too loud, it begins to kill the nerve endings in the inner ear. This hearing loss is called sensorineural hearing loss. As the exposure time to loud noise increases, more and more nerve endings are destroyed. As the number of nerve endings decreases, so does your hearing. There is no way to restore life to dead nerve endings; the damage is permanent.

You can avoid tinnitus by avoiding exposure to loud noises, such as heavy machinery, chain saws, and loud music. When you must expose yourself to loud sounds, use ear protection.

The Decision

Your favorite music group is having a concert this Friday night. Rumor has it that the group is about to break up—this might be your last chance to see them perform. A friend has just purchased tickets for second row seats and is offering to sell one to you for $30. The seat is just a few meters away from the stage speakers. The concert will be about two hours long, and your mother wants you to either wear earplugs or not go at all. What do you do?

Compare the risk assessment and decision-making process you just completed with the method you normally use to make this kind of decision. Would you have reached the same decision? When creating the decision chart, how important was the information on hearing loss? How important were your personal feelings about concerts? Do you think your final decision was largely subjective or objective? Why?

Resources

American Academy of Otolaryngology. March 1991. *Noise, Ears, and Hearing Protection* (pamphlet).
To order, write American Academy of Otolaryngology at:
One Prince Street
Alexandria, VA 22314
Phone: (703) 836-4444
See also Internet site below.

American Speech-Language-Hearing Association. January 1991. *Questions About Noise and Hearing Loss* (pamphlet).
To order, write American Speech-Language-Hearing Association at:
10801 Rockville Pike
Rockville, MD 20852
Phone: (301) 897-5700
See also Internet site below.

Internet sites

American Academy of Otolaryngology
http://www.entnet.org/
Good information and links to other sites.

American Speech-Language-Hearing Association
http://www.asha.org/
Good information and links to other sites.

Hearing Loss Resources
http://www.webcom.com/~houtx/
Contains essays on hearing loss; contact information for an online hearing loss chat group; consumer and medical information; and links to hearing loss resources.

HEARNet
http://www.hearnet.com/
This non-profit organization raises awareness of the dangers of repeated exposure to excessive noise levels.

Smoking

Background

Cigarette smoking is on the rise among high school students. According to the 1995 *Youth Risk Behavior Survey*, more adolescents smoked in 1995 (34.8 percent) than 1993 (30.5 percent) and 1991 (27.5 percent). And the younger the smokers, the more dramatic the increase. In fact, each year one million adolescents smoke their first cigarette. According to the American Lung Association, approximately one-third of these adolescents eventually will die of smoking-related diseases.

Smoking can be both a physical and physiological addiction. Handling a cigarette and the routine of smoking a tobacco product may make people feel more secure. But of greater addictive attraction is the powerful stimulant nicotine. Having an adrenaline-like effect, nicotine increases the heart rate and blood pressure while smoking.

What happens when a cigarette is lit and inhaled? Cigarette smoke contains about 4,000 chemicals, including trace amounts of such toxins as DDT, arsenic, and formaldehyde. In addition, carbon monoxide and other gaseous components enter the smoker's lungs. Carbon monoxide binds to hemoglobin in the blood, taking the place of valuable oxygen. Therefore, smoking decreases the amount of oxygen available to the heart and other parts of the body. Carbon monoxide and other chemicals found in tobacco smoke are also linked to heart disease.

Because the lungs retain 70 to 90 percent of the compounds inhaled from tobacco smoke, the billions of particles from the smoke buildup in the lungs. Tar, for example, contains chemicals that can lead to cancer. If inhaled, tar forms a brown, sticky layer on the linings of the breathing passages.

Advocates of smoking argue that personal choice should be an important factor when it comes to legislating nicotine. They say that outlawing smoking in public areas infringes on personal freedom. The same holds true for taxing tobacco products at a higher rate than other products; many smokers say this so-called "sin tax" is unfair, because it interferes with their personal choice to smoke. Tobacco companies and smoking-rights groups say that second-hand smoke presents only a minor health risk.

The Decision

You are a public policy maker working for the U.S. Surgeon General's Office. The Surgeon General has asked each member of the staff to prepare an analysis of how to reduce the harmful effects of tobacco on citizens of this country.

Possible factors to consider include public health, medical costs, revenue generated from tobacco products, value of a free-market economy, and rights of personal choice. What are the government's options? Which option will you recommend to the Surgeon General?

Write a paragraph or two about how the decision-making process would change if you were making the decision from a personal perspective. How would the goals change? Would outcomes be based more on facts or personal feelings? How would your personal preference (as a smoker or a non-smoker) affect your decision?

Resources

Larson, David E. 1990. *Mayo Clinic Family Healthbook.* New York: William Morrow & Company, Inc.

American Lung Association
Phone: (800) LUNG USA
See also Internet site below.

National Heart, Lung, and Blood Institute
Information Center
P.O. Box 30105
Bethesda, MD 20824-0105
Phone: (301) 251-1222

Internet sites

American Lung Association
http://www.lungusa.org/
Resource for information on tobacco control and lung diseases. Includes information on the 1997 tobacco settlement.

The Smoke Shop
http://www.thesmokeshop.com/
Provides facts and figures about smoking, and links to smokers-rights groups.

STUDENT SECTION

Sources of Energy

Background

Since the Industrial Revolution, the world has become increasingly dependent on technology. This technology, combined with population growth, has led to soaring energy usage. The world's production of commercial energy has grown by 14 percent in the past decade. The world's total energy consumption has increased four-fold since World War II. The task for the next century is to find the best ways of supplying our society with safe, clean, efficient, and reliable energy.

Different countries use different energy resources. Some examples are fossil fuels (oil, coal, and natural gas), nuclear power, electricity, renewable energy (hydropower, geothermal energy, solar thermal energy, photovoltaics, wind power, ocean energy), and biomass (wood or other plant or animal matter that can be burned directly or converted into fuels). Wood, the oldest source of energy, is still the principal fuel for the majority of people in developing countries.

Nuclear power was once hailed as a solution to the energy crisis. But in the United States, nuclear power generates less electrical power than was once anticipated. Nuclear power also has its drawbacks. Nuclear power plants are expensive to build and maintain. They produce radioactive wastes that are dangerous if not properly stored. Furthermore, accidents at Three Mile Island and Chernobyl have created strong public concern about nuclear safety. Despite its drawbacks, nuclear power is relatively clean, produces few emissions, and draws a virtually limitless supply of fuel.

Another energy source, coal, is relatively inexpensive and is in good supply. However, coal has limited applications. Increased coal emissions will also aggravate air quality problems, such as smog and acid rain.

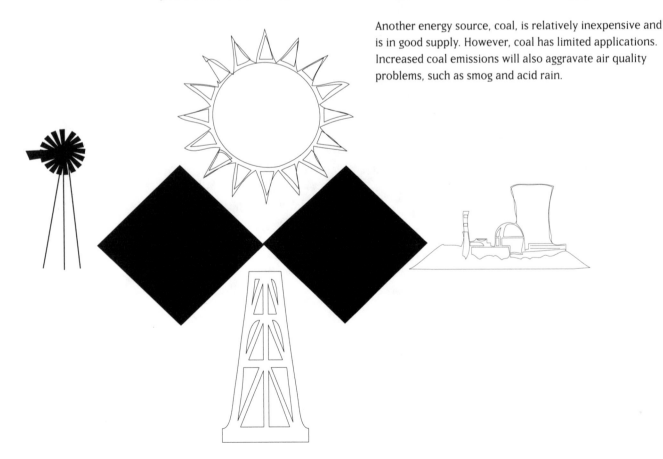

The Decision

Choose the best type of energy source for the future. Keep in mind that each energy source has trade-offs. Research the pros and cons of fossil fuels, nuclear power, and one other energy source. If you have time, strengthen your decision chart by including other alternative energy sources. Is the best solution a single source or a combination of sources?

After you make a decision, discuss it with the rest of the class. What are some possible problems that might occur if the entire world decides to use this approach? What are the benefits? Do you think your final decision is an ideal future energy source? Why or why not?

Resources

Aldridge, Bill, et al. 1996. *Energy Sources and Natural Fuels.* Vol. 2. Arlington, VA: National Science Teachers Association.

Aldridge, Bill, et al. 1993. *Energy Sources and Natural Fuels.* Vol. 1. Arlington, VA: National Science Teachers Association.

Aldridge, Bill, and Crow, Linda. 1995. *Middle Level Energy Series.* Arlington, VA: National Science Teachers Association.

Brown, Lester R., ed. 1996. *State of the World: A Worldwatch Institute Report on Progress Toward a Sustainable Society.* New York: W.W. Norton & Company.

National Science Teachers Association. 1997. *Choices* (poster). Arlington, VA: National Science Teachers Association.

World Resources Institute. 1992. *World Resources, 1992-1993: A Guide to the Global Environment.* New York: Oxford University Press.

Internet sites
Department of Energy
http://www.eia.doe.gov/
Contains information on fuel sources such as petroleum, natural gas, nuclear, coal, and alternative and renewable resources.

Solstice: Center for Renewable Energy and Sustainable Technology
http://solstice.crest.org/docndata.htm
An interesting guide to energy sources such as hydrogen, bioenergy, and geothermal energy.

Humans vs. Robots in Space

Background

In 1957, the Soviet Union surprised the world by launching the first artificial satellite, Sputnik. The launch of Sputnik inaugurated the Space Age in which humanity has extended its reach far beyond Earth. During the next 40 years, humans walked on the moon and spent long periods of time living on orbiting space stations. Robotic spacecraft have explored our nearest neighbors and the far reaches of our Solar System, beaming back spectacular photographs of other planets and their moons.

Throughout the Space Age, supporters of space exploration have engaged in a vigorous debate over the merits of human vs. robotic exploration. Many proponents of human spaceflight view space as a frontier to be explored. They see it as humanity's destiny to explore and eventually undertake mass settlement of space and other worlds. Proponents of human spaceflight believe astronauts are far more capable of exploring this new frontier than robots, which have limited capabilities for movement and thought. This group is not against robotic space exploration, but they prefer to fund programs such as the space shuttle, the international space station, and human flights to the Moon and Mars.

Another group believes that human spaceflight tends to drain money away from more useful robotic exploration efforts. Although robots are not as capable as humans, they are much cheaper to send into space because they don't need food, water, air, and other consumables. Robots can also be sent to areas where humans could not easily survive, such as Venus, where surface temperatures can reach 482 degrees Celsius. Improvements in robotics are gradually narrowing the differences in capability between automated and human exploration. Safety is another important factor; increased use of robotics limits the threat to life that exists in human spaceflight programs.

A frequent target for criticism has been the international space station, which NASA is constructing with Japan, Canada, Russia, and the European Space Agency. Set for completion in 2003, the station will house astronauts who will conduct scientific research and study the effects of long-term human exposure to microgravity. Proponents believe that research in microgravity conditions will lead to breakthroughs in medicine, manufacturing, and other fields, while paving the way for human exploration of the Moon and Mars. Opponents claim the project, which will cost the United States about $60 billion to build and operate, is too expensive. They point out that it is years behind schedule and billions of dollars over budget. Benefits of the program also have been exaggerated, they claim.

The Decision

You are a member of the congressional committee that allocates funding for NASA. Your committee needs to decide whether NASA should continue the space station program. Research how much money NASA is spending on the station and the station's possible benefits. Could money be better used for other programs? How many jobs would be affected by canceling the space station? How many might be generated by spending money on other programs? What would be the impact on our relations with other station partners?

After you have made your decision, analyze it with the rest of the class. What perspective (personal or social) did you use in establishing options, goals, and outcomes? Did other students have the same perspective? Was it difficult to separate viewpoints? Why or why not? From what perspective should this decision be made?

Resources

The Planetary Society
65 North Catalina Avenue
Pasadena, CA 91106-2301
Internet site: http://planetary.org/tps/
Grassroots, pro-space organization supports both human and robotic space exploration, with an emphasis on robotic planetary exploration.

National Space Society
600 Pennsylvania Ave., SE
Washington, DC 20003
Internet site: http://www.nss.org/
Grassroots, pro-space organization dedicated to the creation of a space-faring civilization. Places greater emphasis on human exploration efforts.

Internet sites

NASA
http://www.nasa.gov/

NASA Space Shuttle Launches
http://www.ksc.nasa.gov/shuttle/missions/missions.html/
Visitors to this site learn about current and past National Aeronautics and Space Administration space shuttle launches from the Kennedy Space Center.

1996 National Space Policy—Fact Sheet
National Science and Technology Council
http://www.aiaa.org/policy/nat-space-policy.html/
The Clinton Administration's official policy covers all aspects of space exploration and utilization. Discusses the balance between robotic and human exploration.

Meteors

STUDENT SECTION

Background

Most bodies in the solar system with a solid surface have craters, created by impacts from meteorites, comets, and asteroids. Most asteroids follow circular orbits between the planets of Mars and Jupiter. Sometimes they are found in orbits that cross the paths of Mars or even Earth. Comets follow highly elongated orbits that can come close to Earth or other major bodies. Over the eons, every moon and planet suffers major impacts.

An impact by a small meteorite could be a catastrophe, but would not threaten civilization. (Meteorites are fragments of asteroids large enough to reach the ground.) But an impact by an asteroid larger than one to two kilometers would degrade the global climate, leading to widespread crop failure and loss of life. However, these impacts are highly infrequent. As you can see in Figure I, there is a much greater probability of small meteorites colliding with Earth.

No human in the past 1,000 years is known to have been killed by a meteorite or by the effects of one. NASA knows of no asteroid or comet currently on a collision course with Earth. No large object is likely to strike the Earth in the next several hundred years. Today, astronomers are conducting searches to identify all objects that pose an impact danger to Earth.

FIGURE I:
Probability of
Impact in
Relation to
Meteor Size

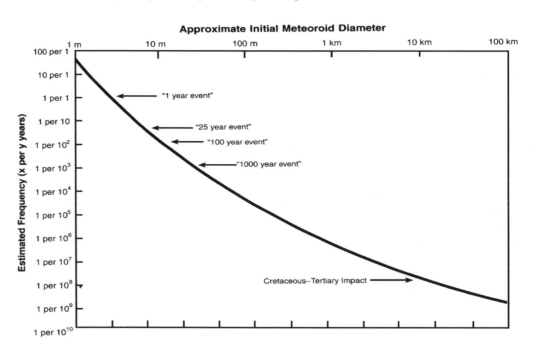

Using sophisticated models that take into account the orbits and gravitational attraction of other planetary bodies, planetary scientists can estimate the likelihood that particular asteroids might come dangerously close to Earth. This likelihood forms the basis of probability estimates given in Figure 1. As you can see in the figure, small impacts are relatively common, while large impacts are fairly rare. *(Note: The Cretaceous-Tertiary Impact refers to an asteroid impact 65 million years ago. Many scientists believe the impact was so destructive that it caused massive extinction of dinosaurs and other species.)*

The Decision

You are a member of the Union of Concerned Scientists (UCS). Your organization conducts research, issues position papers on science policy issues, and frequently testifies before Congress. Imagine that it is 1994, and the collision of Comet Shoemaker-Levy 9 with Jupiter has created great scientific and public concern about large meteorites bombarding Earth. Congress has scheduled hearings on what can be done to prevent this situation. The president of the UCS has appointed you to head a committee to determine the organization's position so it can provide input to Congress.

Research current efforts to improve meteorite and comet detection. Do you think these efforts are adequate? How much would it cost to conduct a more thorough search for Earth-crossing meteorites and comets? Explore whether the U.S. government should develop a defensive system to destroy these bodies or direct them away from the Earth. Is such a system feasible? Consider whether the expenditure of funds for such a defensive system is justified. Develop a decision chart to display your findings, and create a position statement: How do you think the government should address the risk of meteorite impact?

Resources

Hartmann, William K., and Cain, Joe. 1995. *Craters: A Multi-Science Approach to Cratering and Impacts.* Arlington, VA: National Science Teachers Association.

Wright, Russell G. 1997. *Asteroid!* New York: Addison-Wesley.

Internet sites

Asteroid and Comet Impact Hazards
http://george.arc.nasa.gov/sst/
In-depth introduction to asteroid and comet impacts.

NASA
http://www.nasa.gov/

Yahoo Astronomy Site
http://www.yahoo.com/Science/Astronomy
A large astronomy database that can provide the user with information about space probes that can detect Earth-crossing meteorites.

Old-Growth Forests

Background

Everyone agrees that old-growth forests are valuable. But not everyone agrees on how they are valuable, or what to do with them. The commercial logging industry values old-growth forests because they contain very big trees that yield large quantities of good timber. Ecologists value old-growth forests because they provide habitat for many species, play an important role in nutrient cycling, and slow soil erosion. Other groups of people—such as Native Americans, environmentalists, outdoor sports and hunting enthusiasts, and federal and state agencies—value old-growth forests for a broad variety of reasons.

To complicate matters, not everyone agrees on just what an old-growth forest is. How old does a forest have to be before it can be considered old-growth? Many factors, including soil conditions and site qualities, determine the age at which a forest will take on the structural qualities of true old-growth. Scientists generally agree that a true old-growth forest is between 250 and 350 years old. But there is some evidence that old-growth structural qualities will appear in a forest after only 80 to 100 years. Loggers prefer the 80 to 100-year definition because it means a shorter cycle for planting, harvesting, and marketing trees.

Some of the last remaining stands of old-growth forests in the United States are in the Pacific Northwest, especially Washington State's Olympic Peninsula. Many communities in the Pacific Northwest rely on commercial logging for their livelihoods. For this and other reasons, there has been much debate about what to do with old-growth forests. Can the needs of local and regional economies be balanced with the ecological need to conserve such a valuable natural resource?

The Decision

An area of old-growth forest in the Pacific Northwest is for sale. Because so many different groups of people have so many different ideas about what should be done with it, the government has stepped in to arbitrate the decision-making process. Your teacher will divide the class into several smaller groups. Each group will be asked whom they wish to represent in deciding what to do with this forest. Possibilities for group roles include loggers, ecologists, hunters, environmentalists, local citizens, and government officials. While your group can choose any role it wants, make sure that each group has a different role to play in deciding what to do with this old-growth forest.

As your group discusses its decision, one or more of you should take notes about your decision-making process. Decide within your group what your objectives are. Do they conflict with the objectives of other groups? Does your group have any objectives in common with other groups? If there are common or similar objectives, how might you use this in your group's decision-making process?

After developing your group's decision, rejoin the class. Leave the role you are playing, and think of yourself as an impartial observer. From this viewpoint, try to develop an unbiased decision chart that accurately represents the work of the entire class. How did your assigned group roles affect your decisions? Did the role affect your ability to be objective? Was the "impartial observer" better able to make an objective decision? Is it possible to make a decision that is completely objective?

Resources

National Science Teachers Association. 1996. *Biodiversity*. Global Environmental Change Series. Arlington, VA: National Science Teachers Association.

National Science Teachers Association. 1997. *Deforestation*. Global Environmental Change Series. Arlington, VA: National Science Teachers Association.

Botkin, D. B. 1993. *Forest Dynamics: An Ecological Model*. New York: Oxford University Press.

American Forest and Paper Association
1111 19th St. NW, Suite 800
Washington, DC 20036
Phone: (202) 463-2700
Internet site: http://www.afandpa.org/

The Nature Conservancy
1815 Lynn St.
Arlington, VA 22209
Phone: (800) 628-6860
Internet site: http://www.tnc.org/

U.S. Forest Service
P.O. Box 96090
Washington, DC 20090
Phone: (202) 205-1760
Internet site: http://www.fs.fed.us/

Internet sites

Earthviewer
http://www.fourmilab.ch/earthview/vplanet.html
Allows students to view their local area from a satellite.

Environmental Organizations Web Directory
http://www.webdirectory.com/
A search engine to connect the user with a wide variety of environmental groups.

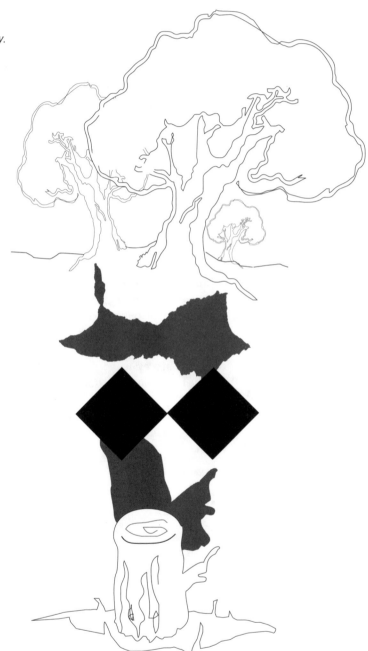

Severe Weather

Background

Perfect weather is frequently described in poetry and songs as being gentle breezes, a gentle evening rain, or blue skies. People have different temperatures and weather ranges they prefer; and when the air is hotter or colder, the variation may encourage conversations.

Climates in the U.S. include low desert, high desert, subtropics, tundra, and temperate zones. Blizzards, tornadoes, hurricanes, droughts, and flooding can occur in many locations, but these types of severe weather are more common in some areas than others. But even Death Valley, California occasionally experiences flooding and Miami, Florida sometimes receives heavy snow. In 1997, flooding along the Red River heavily damaged North Dakota communities that hadn't experienced such high water since they had been built. These unusual weather occurrences are labeled "50 year or 100 year events." Such a designation doesn't mean that unusual events will come along every 50 years; it means that statistically, an event would be expected to occur only once in every 50 years. For example, an event could happen twice in a decade—then 90 years could pass before it happens again.

Because weather is so important to our lives, governments around the world spend considerable effort and money to monitor it. The collected data, gathered by thousands of weather stations around the globe, are placed in huge databases and used in a variety of ways. Weather broadcasters use these data when they announce that this was the hottest July 4th in your community since 1957. Meteorologists can estimate weather conditions on a given day and discover whether it rained under similar conditions in previous years. Meteorologists also can use the data to produce climate models that provide better long-range forecasts.

Many climate databases are available on the Internet. A few sites are listed in the resources for this exercise; you may well find others to answer your questions about weather patterns and severe weather.

The Decision

Put yourself in the role of a farmer looking to purchase a new farm. You have found fertile and affordable land near the cities listed on the worksheet on the next page. But severe weather is especially tough on crops and livestock, and you want to choose a location that poses the least weather danger to your investment.

The worksheet includes information on severe weather. Determine the weather quality in your community by using the Internet, local newspapers and television stations, or other resources. If there is other weather data you would like to consider in locating your farm, extend the worksheet to include it. (You can also include other potential benefits and risks of building a farm in these locations.) Complete the worksheet, and decide which location will be best for your new farm.

	Tornadoes/yr.	Hurricanes/yr.	Blizzards/yr.	Floods/yr.
Fargo, ND	0	0	2	0.1
Charleston, SC	1	0.2	0	4
Tulsa, OK	44.2	0	0	1
Your hometown				

WORKSHEET:
Severe Weather
Comparisons

Resources

Smith, Sean P., and Ford, Brent A. 1994. *Project Earth Science: Meteorology.* Arlington, VA: National Science Teachers Association.

National Geographic Society
Education Services
17th and M Streets, NW
Washington, DC 20036
Phone: (202) 921-1330
Internet site: http://www.nationalgeographic.com/

National Center for Atmospheric Research
Information and Education Outreach Program
P.O. Box 3000
Boulder, CO 80307-3000
Phone: (303) 497-8600
Internet site: http://www.ncar.ucar.edu/

Internet sites
NOAA
http://www.noaa.gov/
National Climatic Data Center
http://www.ncdc.noaa.gov
A good starting place for atmospheric research.

National Severe Storms Laboratory
http://www.nssl.uoknor.edu/
Contains raw data for everything you will ever want to know about tornadoes in the United States.

The Weather Page
http://www.esdim.noaa.gov:80/weather_page.html

Weather Underground: The University of Michigan
http://blueskies.sprl.umich.edu/
Here you will find access to weather imagery and current conditions and forecasts.

Floodplains

Background

In 1993, disastrous flooding along the Mississippi River focused attention on federal programs for reducing flood damage. A presidential task force created after the flood called for less reliance on levees and better management of the nation's flood plains and river basins. This Interagency Floodplain Management Review Committee called for relocation of vulnerable homes and businesses; managing floodplains as components of larger watersheds; and linking future disaster relief to good flood plain and watershed management.

The task force called for an overhaul of the Army Corps of Engineer's planning regulations, the enactment of a National Floodplain Management Act to better coordinate federal agencies, and the creation of basin-wide planning commissions to develop and implement ecosystem management strategies. It was found that flood damage can be minimized by controlling runoff, planning the use of the land, and identifying areas at risk. The report noted that the loss of wetlands, upland vegetation and other modifications of the landscape had significantly increased runoff, and called for the restoration of wetlands, and better land management practices in order to alleviate the flooding. The task force also found that levees often make flooding worse.

The Federal Emergency Management Agency (FEMA) works with state and community governments to identify their flood hazards areas. The National Flood Insurance Program is another national program to alleviate flood damage. The program enables property owners to buy flood insurance at reasonable rates. In return, communities carry out local flood plain management measures to protect lives and property from future flooding. The communities must require permits for construction in flood plains and ensure that construction will minimize flood damage.

The Flood of 1993

♦ Record rains drenched the upper Mississippi River basin, with over one meter of rain from April to July 1993. It was the worst flooding in 30 years.

♦ Areas in the states of Minnesota, Iowa, Missouri, and Illinois were flooded. Over nine million hectares were flooded.

♦ Levees were built to protect cities, villages, and farms along the river. But levees may do more harm than good because they cause flood waters to swell upstream and go over the tops of levees. They also cause water to run faster and deeper, eroding levees downstream.

♦ The flood caused more than $10 billion worth of damage to crops and property, resulted in 50 deaths, destroyed 72,000 homes, and affected 580 hectares of land.

♦ The town of Grafton, Illinois, is planning on moving to higher ground. It has been flooded six times in 20 years. Nearly 40 percent of the towns' homes and businesses would qualify to receive federal aid. The total cost of the move would be $25 million.

The Decision

You are a resident of a small town in Missouri. Your town was nearly destroyed during the 1993 flood. Several people were killed and the loss of property was staggering. Should you move your town? The town has been located on the Mississippi River for over 100 years and is rich in history. Make a list of the town's possible options. Add to the list some hypothetical options for your own family. Which option is best? Are your family's needs best met by the option that is best for the town?

Resources

Wright, Russell G. 1996. *Flood!* New York: Addison-Wesley.

American Geophysical Union
2000 Florida Ave., NW
Washington, DC 20009
Phone: (202) 462-6900
Internet site: http://earth.agu.org/

Federal Emergency Management Agency (FEMA)
Federal Center Plaza
500 C Street, SW
Washington, DC 20472
Internet site: http://www.fema.gov/
See also http://www.fema.gov/fema/flood.html/

Internet sites

EIS International: Hazard Technology
http://www.eisintl.com/wtshapng/hastech/art24.html

Oahu Civil Defense Agency
http://www.hgea.org/E911/mnge.htm/
Reading: "The Community's Role in Floodplain Management"

WWW Virtual Library: Hazards and Risk, Meteorological—Floods (riverine)
http://life.csu.edu.au/hazards/5Riverine.html/
Connects to many flood-related sites.

Curriculum Matrices

APPENDIX A

Use the matrices on the following pages to connect the activities in the book with your current curriculum and the *National Science Education Standards*. The column headers identify four content categories and outline the activities' main curriculum connections: physical, life, Earth and space sciences, and social perspectives. Under each heading are the fundamental concepts and principles pertaining to the specific activity. For detailed descriptions of these fundamental concepts, see the *National Science Education Standards* and *NSTA Pathways to the Science Standards, High School Edition* (in Selected Resources).

Four content categories (Unifying Concepts and Processes, Science as Inquiry, Science and Technology, and History and Nature of Science) are not included in the matrices because all the decision-making activities cover these standards. Specifically, the relevant fundamental concepts are: Systems, order, and organization; Evidence, models, and explanation; Understandings about scientific inquiry; Understandings about science and technology; Science as a human endeavor; and Nature of scientific knowledge. Again, see the *Standards* for more detailed descriptions.

PART TWO: GUIDED ACTIVITIES

ACTIVITY	Physical Science	Life Science	Earth and Space Science	Science in Personal and Social Perspectives
Xenotransplants		• The cell • Biological evolution • Interdependence of organisms • Behavior of organisms		• Personal and community health • Natural and human-induced hazards • Science and technology in local, national, and global challenges
Immunizations		• The cell • Biological evolution • Matter, energy, and organization in living systems		• Personal and community health • Natural and human-induced hazards • Science and technology to local, national, and global challenges
Household Cleaning Products	• Structure and properties of matter • Chemical reactions	• The cell		• Personal and community health • Environmental quality • Natural and human-induced hazards
Ozone	• Structure of atoms • Structure and properties of matter • Chemical reactions • Interactions of energy and matter		• Energy in the Earth system • Geochemical cycles	• Environmental quality • Natural and human-induced hazards
Groundwater	• Chemical reactions • Motions and forces		• Energy in the Earth system • Geochemical cycles • Origin and evolution of the Earth system	• Personal and community health • Natural resources • Environmental quality • Natural and human-induced hazards
The Politics of Biodiversity		• Biological evolution • Interdependence of organisms • Matter, energy, and organization in living systems	• Geochemical cycles	• Natural resources • Environmental quality
Speed Limits	• Motions and forces • Conservation of energy and increase in disorder			• Personal and community health • Natural and human-induced hazards
Roller Coasters	• Motions and forces • Conservation of energy and increase in disorder			• Personal and community health • Natural and human-induced hazards
Recycling	• Conservation of energy and increase in disorder		• Geochemical cycles	• Natural resources • Environmental quality • Natural and human-induced hazards • Science and technology in local, national, and global challenges
A Local Decision	Topic depends on local issues	Topic depends on local issues	Topic depends on local issues	• Personal and community health • Science and technology in local, national, and global challenges

PART THREE: INDEPENDENT EXERCISES

ACTIVITY	Physical Science	Life Science	Earth and Space Science	Science in Personal and Social Perspectives
Bovine Growth Hormone	• Chemical reactions	• The cell • Matter, energy, and organization in living systems • Behavior of organisms		• Personal and community health • Natural and human-induced hazards
Chemical Warfare	• Structure of atoms • Structure and properties of matter • Chemical reactions	• The cell		• Natural and human-induced hazards • Science and technology in local, national, and global challenges
Asbestos	• Structure and properties of matter	• The cell		• Personal and community health • Natural and human-induced hazards
Zoos		• Biological evolution • Interdependence of organisms • Population growth • Behavior of organisms		• Natural resources • Science and technology in local, national, and global challenges
Marine Resources		• Biological evolution • Interdependence of organisms • Matter, energy, and organization in living systems		• Natural resources • Environmental quality • Science and technology in local, national, and global challenges
Diet Decisions		• The cell • Matter, energy, and organization in living systems		• Personal and community health • Natural and human-induced hazards
Hearing Loss	• Motions and forces • Interactions of energy and matter	• Behavior of organisms		• Personal and community health • Natural and human-induced hazards
Smoking	• Chemical reactions	• The cell		• Personal and community health • Natural and human-induced hazards
Sources of Energy	• Chemical reactions • Motions and forces • Conservation of energy and increase in disorder • Interactions of energy and matter		• Energy in the Earth system • Geochemical cycles	• Natural resources • Environmental quality • Science and technology in local, national, and global challenges
Humans vs. Robots in Space		• Behavior of organisms	• Origin and evolution of the Earth system • Origin and evolution of the universe	• Science and technology in local, national, and global challenges

(Continued on next page)

PART THREE (continued)

ACTIVITY	Physical Science	Life Science	Earth and Space Science	Science in Personal and Social Perspectives
Meteors			• Energy in the Earth system • Origin and evolution of the Earth system • Origin and evolution of the universe	• Natural and human-induced hazards
Old-Growth Forests		• Biological evolution • Interdependence of organisms • Matter, energy, and organization in living systems	• Geochemical cycles	• Natural resources • Environmental quality
Severe Weather	• Motions and forces		• Energy in the Earth system	• Natural and human-induced hazards
Floodplains	• Conservation of energy and increase in disorder		• Energy in the Earth system • Origin and evolution of the Earth system	• Environmental quality • Natural and human-induced hazards

Summary of Decision Making

APPENDIX B

STEP ONE: What's the Decision?

(A) Before you do any analysis, write a tentative decision in your journal.

(B) Whose decision is it? Determine at what level (personal, local, national, international) the decision is being made, and make sure all goals and options are stated at this level.

(C) Define the hazard and evaluate its risk (intensity of effects and exposure). Consider risk space and human health, ecological, economic, governmental, and ethical effects. Also, consider effects in present and future. Be as specific as possible, but leave "holes" if a lack of information is holding up the process. Return to the research later; or decide without the information.

(D) Use the risk assessment to make a list of goals (values) for the decision. The list of goals is a preliminary list which can be amended.

STEP TWO: What Should Happen?

(A) Create a decision chart using the goals from Step One.

(B) Deliberate on and select two to four options.

(C) Enter best guesses on outcomes.

(D) Identify initial research needs from outcome boxes that are missing information.

STEP THREE: What Do We Know?

(A) Research outcomes.

(B) Make probability estimates. Keep in mind that events are either all-or-nothing or continuum.

(C) Complete decision chart with information from research.

STEP FOUR: What's the Answer?

Decision chart analysis

(A) Method One: Draw importance bars based on the advantages of options for each goal. Add together the importance bars. Which option is the winner?

(B) Method Two: Assign numerical values and probabilities for each outcome in the chart. Compute expected value. Which option is the winner?

(C) Sensitivity analysis: Check the uncertainty of outcomes to identify research needed.

Final decision: Any additional issues?

(A) Are you happy with the final decision? Why or why not? Were there missing factors?

(B) What research is needed to make the decision more conclusive? Identify the single most important fact.

Blank Decision Chart

APPENDIX
C

	OPTION 1:	OPTION 2:	OPTION 3:
GOAL A:			
GOAL B:			
GOAL C:			

Worst   **IMPORTANCE BARS**   Best

Goal A

Goal B

Goal C

Assessment Rubrics

APPENDIX
D

Assessment Steps

When students make decisions based on science, they apply facts and science tools they have mastered. Measuring student ability to apply learning—to reach decisions—is itself a useful method to assess science coursework, and can be incorporated as a culminating exercise among many areas of science instruction. Decision making is not an extra topic to fit into your curriculum, but a way of ensuring students understand the applications of science concepts. For example, see Appendix A to connect the activities to specific curriculum topics.

This appendix provides assessment rubrics to use to evaluate the process of, and concepts behind, decision making. You may wish to use these rubrics as a guide to planning instruction.

The first step in the assessment process is to develop learning objectives. Suggested objectives are given in the assessment rubrics. However, you may also develop your own objectives. Or, involve your students: As they do the decision-making process, have them write their own ideas for what they should be learning.

Students will use the Summary of Decision Making (page 119) as a guide as they do activities in this book. The summary also works as a procedural self-assessment tool. Students can see exactly the steps they should complete.

If your students keep a journal of their science learning (as suggested on page xi), the journal will provide a summary of material for students to draw from for their decision-making process, as well as a place to record how they made their decisions. As you assess student journals, make sure you give feedback on areas that need improvement.

For additional assessment ideas, see *NSTA Pathways to the Science Standards, High School Edition* in Selected Resources.

How to Use the Rubrics

Two assessment rubrics give you some flexibility in reviewing student work. The procedural assessment rubric is based on the decision-making process. Students are assessed on how they complete all the steps. The concept assessment rubric is based on general decision-making skills.

The easiest way to use the rubrics is to give points for each row's skill. Total the scores for the column and assign a grade. For example, the maximum points for the concept assessment would be 24 (*i.e.*, three points multiplied by eight categories).

Another method is to determine the column in which the majority of the student's work fits. The column with the most answers is the descriptive evaluation (*i.e.*, "Overall Excellent"). Then, give students some feedback on the specific areas where they could improve.

Process Assessment

	EXCELLENT (3 points)	OK (2 points)	NEEDS WORK (0 or 1 point)
STEP ONE			
Whose decision is it?	Determines decision level. Gives good explanation of why decision is best made at that level. All goals and options reflect the level at which decision is made.	Determines decision level. Some goals and options reflect those of identified decision maker.	Confuses decision makers. Goals and options do not reflect those of identified decision maker.
Risk assessment	Defines the hazard and evaluates its risk (intensity of effects and exposure). Considers risk space and human health, ecological, economic, governmental, and ethical effects. Considers present and future effects.	Defines the hazard and evaluates risks, but leaves out some categories of effects. Considers present and future effects.	Defines hazard but doesn't evaluate risk completely. Doesn't assess effects in both present and future.
Identifies goals	Uses risk assessment to make a list of pertinent goals (values) for the decision.	Makes a list of goals (values) for the decision.	Identifies goals that are not pertinent to the decision.
STEP TWO			
Create decision chart	Creates decision chart in proper format, using the goals.	Creates decision chart.	Creates chart in incorrect format.
Select options	Selects particularly good options, which are not a combination of actions and which show some thought.	Selects two to four pertinent options.	Selects options that are unrealistic or a combination of actions.
Estimate outcomes	Enters outcomes using class knowledge in chart. Identifies initial research needs from outcome boxes missing information.	Enters best guesses on outcomes in chart. Identifies initial research needs from outcome boxes missing information.	Enters outcomes. Doesn't identify research needs.

	EXCELLENT (3 points)	OK (2 points)	NEEDS WORK (0 or 1 point)
STEP THREE			
Research	Researches all outcomes in chart and identifies problem outcomes (uncertain, unknown). Gives reasons for uncertainty.	Researches outcomes.	Leaves some outcomes unresearched/unexplained.
Probability estimates	Makes probability estimates using numbers for both all-or-nothing and continuum events.	Makes probability estimates using numbers, but doesn't distinguish between two kinds of events.	Makes a few estimates using words only. Doesn't distinguish between two kinds of events.
Complete decision chart	Completes decision chart with information from research and original thought. For blank boxes, states the uncertainty and gives explanation.	Completes decision chart with information from research.	Leaves some boxes blank with no explanation.
STEP FOUR			
Analysis	Draws importance bars. Uses bars and expected value for analysis.	Draws importance bars and uses them for analysis. Does not use expected value.	Does little analysis based on importance bars or expected value. Gives "gut feeling" not based on previous research.
Sensitivity analysis	Checks each outcome for sensitivity. Identifies further research needed.	Checks a few outcomes for sensitivity.	Doesn't do sensitivity analysis.
Final decision: Any additional issues?	Discusses process. Identifies missing factors. Identifies research needed to make the decision more conclusive. Identifies the main obstacle to making decision.	Discusses process. Identifies some additional research needs.	Reaches conclusion but doesn't explain process. Doesn't identify any additional research needed.

Concept Assessment

SKILL	EXCELLENT (3 points)	OK (2 points)	NEEDS WORK (0 or 1 point)
Research	*Does significant outside research from a variety of sources.*	*Does some limited research.*	*Does minimal or no research.*
Science application	*Applies all pertinent knowledge from curriculum and outside research to decision.*	*Applies knowledge from science curriculum to decision.*	*Doesn't demonstrate class knowledge in decision.*
Original thought	*Demonstrates significant insightful and original thought about the issue.*	*Completes decision. Gives answers based on logic.*	*Doesn't think through decision. Demonstrates no original thought.*
Identifying uncertainties	*Identifies most uncertainties and suggests research to refine analysis.*	*Identifies some areas of uncertainty.*	*Doesn't identify areas of uncertainty.*
Analysis	*Uses importance bars and sensitivity analysis to analyze decision.*	*Uses importance bars to analyze decision.*	*No analysis. Uses gut feelings.*
Final decision summary	*Makes conclusion using importance bars and expected value. Compares final decision to one not based on analysis.*	*Makes conclusion based on importance bars. Identifies final decision.*	*Makes decision, but doesn't explain process.*
Writing	*Writes persuasively and clearly. Uses logical arguments to summarize position.*	*Writes clearly. Explains the decision situation.*	*Writes a bit unclearly. Doesn't fully explain the thought process involved in decision making.*
Math applications	*Uses math to compute expected value and express probabilities.*	*Uses math to express probabilities.*	*Doesn't use math.*

Selected Resources

RESOURCES

Decision Analysis and Risk Assessment

Keeney, Ralph. 1992. *Value-Focused Thinking: A Path to Creative Decision Making.* Cambridge, MA: Harvard University Press.

Kunreuther, Howard, and Slovic, Paul, eds. 1996. *The Annals of the Academy of Political and Social Science: Challenges in Risk Assessment and Risk Management* 545 (May).

Lindley, Dennis V. 1985. *Making Decisions.* 2nd ed. New York: Wiley.

Stern, Paul, and Fineberg, Harvey, eds. 1996. *Understanding Risk: Informing Decisions in a Democratic Society.* Washington, DC: National Academy Press.

Watson, Stephen, and Buede, Dennis. 1987. *Decision Synthesis: The Principles and Practice of Decision Analysis.* New York: Cambridge University Press.

National Research Council. 1996. *National Science Education Standards.* Washington, DC: National Academy Press.

National Science Teachers Association. 1996. *NSTA Pathways to the Science Standards.* High School Edition. Arlington, VA: National Science Teachers Association.
Connects the Standards *to current issues and science curricula. See assessment section which includes ideas on using student writing and portfolios.*

National Science Teachers Association. 1990. *Real Science, Real Decisions: A Collection of Thinking Activities from* The Science Teacher. Arlington, VA: National Science Teachers Association.

Tierney, Robert. 1996. *How to Write to Learn Science.* Arlington, VA: National Science Teachers Association.
Includes ideas for assessing student writing and ways to improve notetaking and journal writing skills.

Teaching Resources

Foundation for American Communications and National Sea Grant College Program. 1995. *Reporting on Risk: A Handbook for Journalists and Citizens.* Annapolis, MD: The Annapolis Center.
Appropriate for high school students. Excellent summary of risk assessment principles. Recommended for evaluating studies in the news.

LOGAL.net software
This software allows teachers to access science and math activities on-line. For more information, contact:
LOGAL
125 Cambridge Park Drive
Cambridge, MA 02140
(617) 491-4440
http://www.logal.com

Periodicals

These periodicals cover developing issues in science. They are a good place for students to start their research. Some also have on-line versions (see Internet Sites).
Nature
New Scientist
Science
Science News
Scientific American
The Science Teacher
Quantum

Internet Sites

These are general resources for background on an assortment of issues. See activities for sites pertaining to a specific issue.

Decision Analysis Society
http://www.fuqua.duke.edu/faculty/daweb/
A Lexicon of Decision Making
http://www.fuqua.duke.edu/faculty/daweb/lexicon.htm

ERIC Clearinghouse for Science, Mathematics, and Environmental Education
http://www.ericse.org/

ERIC Clearinghouse for Social Studies/Social Science Education
http://www.indiana.edu/~ssdc/eric_chess.htm
Current Events Link
http://www.indiana.edu/~ssdc/curlinks.html
Provides links to on-line news networks and popular news magazines.

Nature on-line
http://www.nature.com/

New Scientist on-line
http://www.newscientist.com/

Science on-line
http://www.sciencemag.org/

Scientific American on-line
http://www.sciam.com/

Science Learning Network
http://www.sln.org/
Sponsored by the National Science Foundation.

Why Files: Science Behind the News
http://whyfiles.news.wisc.edu/
Sponsored by the National Institute for Science Education. Gives easy-to-understand background on current newsworthy topics.

U.S. Government Internet Sites

The following provides an overview of useful internet sites of major U.S. government branches, departments, agencies, and commissions.

The White House
http://www.whitehouse.gov/WH/Welcome.html

Congress
http://congress.org/

Departments
Agriculture
http://www.usda.gov/

Commerce
http://www.doc.gov/

National Oceanic and Atmospheric Administration
http://www.noaa.gov/

Defense
http://www.dtic.dla.mil/defenselink

Education
http://www.ed.gov/

Energy
http://www.doe.gov/

Health and Human Services
http://www.os.dhhs.gov/

    Centers for Disease Control and Prevention
    http://www.cdc.gov/

    Food and Drug Administration
    http://www.fda.gov/

    National Institutes of Health
    http://www.nih.gov/

Housing and Urban Development
http://www.hud.gov/

Interior
http://www.doi.gov/

    U.S. Geological Survey
    http://www.usgs.gov/

Justice
http://www.usdoj.gov/

Labor
http://www.dol.gov/

State
http://www.state.gov/

Transportation
http://www.dot.gov/

Treasury
http://www.ustreas.gov/

Veterans Affairs
http://www.va.gov/

Federal Agencies and Commissions
Consumer Product Safety Commission
http://cpsc.gov/

Environmental Protection Agency
http://www.epa.gov/

National Aeronautics and Space Administration
http://www.nasa.gov/

National Science Foundation
http://www.nsf.gov/

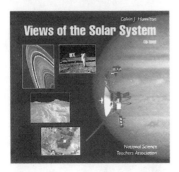